Food Analysis Laboratory Manual

edited by

S. Suzanne Nielsen

Purdue University
West Lafayette, Indiana

KLUWER ACADEMIC / PLENUM PUBLISHERS
New York • Boston • Dordrecht • London • Moscow

Library of Congress Cataloging-in-Publication Data

Food analysis laboratory manual/edited by S. Suzanne Nielsen.
    p. cm. — (Food science text series)
    "Written to accompany the textbook, Food Analysis, third ed."
    Includes bibliographical references and index.
    ISBN 0-306-47496-4
     1. Food—Analysis—Handbooks, manuals, etc. I. Food analysis. II. Title. III. Series.

TX545 .N47 2003
664'.07—dc21

2002042761

ISBN: 0-306-47496-4

©2003 Kluwer Academic / Plenum Publishers, New York
233 Spring Street, New York, New York 10013

http://www.wkap.com

10 9 8 7 6 5 4 3 2 1

A C.I.P. record for this book is available from the Library of Congress

Permissions for books published in Europe: *permissions@wkap.nl*
Permissions for books published in the United States of America: *permissions@wkap.com*

Printed in the United States of America

Contents

Preface and Acknowledgments

This laboratory manual was written to accompany the textbook, *Food Analysis*, third edition. The laboratory exercises are tied closely to the text, and cover 19 of the 32 chapters in the textbook. Most of the laboratory exercises include the following: background, reading assignment, objective, principle of method, chemical (with CAS number and hazards), reagents, precautions and waste disposal, supplies, equipment, procedure, data and calculations, questions, and references.

Instructors using these laboratory exercises should note the following:

1. It is recognized that the time and equipment available for teaching food analysis laboratory sessions varies considerably between schools, as do student numbers and their level in school. Therefore, instructors may need to modify the laboratory procedures (e.g., number of samples analyzed; replicates) to fit their needs and situation.
2. The information on hazards and precautions in use of the chemicals for each experiment is not comprehensive, but should make students and a laboratory assistant aware of major concerns in handling and disposal of the chemicals.
3. It is recommended in the text of the experiments that a laboratory assistant prepare many of the reagents, because of the time limitations for students in a laboratory session. The lists of supplies and equipment for experiments do not necessarily include those needed by the laboratory assistant in preparing reagents, etc. for the laboratory session.
4. The data and calculations section of the laboratory exercises provides details on recording data and doing calculations. In requesting laboratory reports from students, instructors will need to specify if they require just sample or all calculations.
5. Students should be referred to the definitions on percent solutions and on converting parts per million solutions to other units of concentration as given in the notes that follow the preface.

With this being the first edition of a laboratory manual, there are sure to be inadvertent omission and mistakes. I will very much appreciate receiving from instructors, with input from lab assistants and students, suggestions for revisions.

I am grateful to the food analysis instructors identified in the text who provided complete laboratory experiments or the materials to develop the experiments. I especially thank Dr Charles Carpenter of Utah State University for sharing his ideas about content of the experiments. I also want to thank the following students and others from my laboratory group who helped work out and test the experimental procedures written: Susan Axel Bedsaul, Kirby Hayes, Christina Kokini, Noël Davis Larson, Carla Mjier, Carol Rainford, and Katie Rippel. Additional thanks are extended to Susan Axel Bedsaul and Christina Kokini for their work in editing, word processing, and checking on details of the experiments.

S. Suzanne Nielsen

Notes on Calculations of Concentration

Definitions of Percent Solutions:

Weight/Volume Percent (%, w/v)

    = weight, in g of a solute, per 100 ml of solution

Weight/Weight Percent (%, w/w)

    = weight, in g of a solute, per 100 g of solution

Volume/Volume Percent (%, v/v)

    = volume, in ml of a solute, per 100 ml of solution

Concentration of minerals is expressed commonly as parts per billion (ppb) or parts per million (ppm). Parts per million is related to other units of measure as follows:

$$\text{ppm} = \frac{\mu g}{g} = \frac{mg}{1000\ g} = \frac{mg}{L}$$

$$1000\ \text{ppm} = \frac{1000\ \mu g}{g} = \frac{1\ mg}{g} = \frac{0.001\ g}{g}$$

$$= \frac{0.1\ g}{100\ g} = 0.1\%$$

Nutritional Labeling Using a Computer Program

Laboratory Developed by

Dr Lloyd E. Metzger,
University of Minnesota, Department of Food Science and Nutrition, St. Paul,
Minnesota

INTRODUCTION

Background

The 1990 Nutrition Labeling and Education Act mandated nutritional labeling of most foods. As a result, a large portion of food analysis is performed for nutritional labeling purposes. A food labeling guide and links to the complete nutritional labeling regulations are available on-line at http://vm.cfsan.fda.gov/~dms/flg-toc.html. However, interpretation of these regulations and the appropriate usage of rounding rules, available nutrient content claims, reference amounts, and serving size can be difficult.

Additionally, during the product development process the effect of formulation changes on the nutritional label may be important. As an example, a small change in the amount of an ingredient may determine if a product can be labeled low fat. As a result, the ability to immediately approximate how a formulation change will impact the nutritional label can be valuable. In some cases, the opposite situation may occur and a concept called reverse engineering is used. In reverse engineering the information from the nutritional label is used to determine a formula for the product. Caution must be used during reverse engineering. In most cases, only an approximate formula can be obtained and additional information not provided by the nutritional label may be necessary.

The use of nutrient databases and computer programs designed for preparing and analyzing nutritional labels can be valuable in all of the situations described above. In this laboratory you will use a computer program to prepare a nutritional label from a product formula, determine how changes in the formula affect the nutritional label, and observe an example of reverse engineering.

Reading Assignment

Nielsen, S.S., and L.E. Metzger. 2003. Nutrition labeling. Ch. 3, in *Food Analysis*, 3rd ed. S.S. Nielsen (Ed.), Kluwer Academic, New York.
Owl Software. 2002. TechWizard™ Version 3 Manual, LaVic, Inc., Lancaster, PA.

Objective

Prepare a nutritional label for a yogurt formula, determine how formulation changes will affect the nutritional label, and observe an example of reverse engineering.

Materials

TechWizard™ Version 3: Formulation and Nutrition Labeling Software PC with Microsoft® Excel 97 or Excel 2000

Notes

Instructions on how to receive and install the software used for this laboratory are located on-line at **www.owlsoft.com**. On the left hand side of the web page click on the *Food Analysis Students* link located under the services heading. **It is possible that the TechWizard™ program has been updated since the publication of this laboratory manual and any changes in the procedures described below will also be found on this web page.**

*Install the software prior to the laboratory to ensure that it is working properly with your PC.

METHOD A: PREPARING NUTRITION LABELS FOR SAMPLE YOGURT FORMULAS

Procedure

1. Start the TechWizard™ program (*Click on the Enable Macros button*) and select Enter Program (*Click on OK*). Enter the Nutrition Labeling section of the program (*From the Labeling menu select View Nutrition Labeling Section and Nutrition Formula*).
2. Enter the ingredients for formula #1 listed in Table 1-1. (*Click on the Add Ingredients button, then select each ingredient from the ingredient list window and click on the Add button, click on the X to close the window after all ingredients have been added.*)
3. Enter the percentage of each ingredient for formula #1 in the % (wt/wt) column—selecting

Sample Yogurt Formulas

	Formula #1 (%)	Formula #2 (%)
Milk (3.7% fat)	38.201	48.201
Skim milk no Vit A add	35.706	25.706
Condensed skim milk (35% total solids)	12.888	12.888
Sweetener, sugar liquid	11.905	11.905
Modified starch	0.800	0.800
Stabilizer, gelatin	0.500	0.500

the Sort button above that column will sort the ingredients by the % (wt/wt) in the formula.

4. Enter the serving size (common household unit and the equivalent metric quantity) and number of servings. (*First click on the Serving Size button under Common Household unit, enter 8 in the window, click on OK, select oz from the units drop down list; next click on the Serving Size button under Equivalent Metric Quantity, enter 227 in the window, click on OK, select g from the units drop down list; and finally click on the Number of Servings button, enter 1 in the window, click on OK.*)

   *Note by clicking on the Show Ref. Table button, a summary of the CFR 101.12 Table 2 Reference Amounts Customarily Consumed Per Eating Occasion will be displayed.

5. Enter a name and save formula #1. (*Click on the Formula Name window, enter "food analysis formula #1" in the top Formula Name window, click on the X to close the window, from the File menu select Save Formula.*)

6. View the nutrition label and select label options (*click on the View Label button, click on the Label Options button, select the label type you want to display*—the standard, tabular, linear, or simplified format can be displayed; *select the voluntary nutrients you want to declare*—you may want to select *Protein—Show ADV* since yogurt is high in protein; the daily value footnote and calories conversion chart will be displayed unless *Hide Footnote* and *Hide Calorie Conversion Chart* are selected; *when you have finished selecting the label options select Apply and then Close to view the label*).

7. Edit the ingredient declarations list (*Click on the View/Edit Declaration button, click Yes when asked*—Do you wish to generate a formula declaration using individual ingredient declarations?—*Each ingredient used in formula can be selected in the top window and the ingredient declaration can be edited in the middle window.*)

   *Note the rules for ingredient declaration are found in the CFR 101.4.

8. Print the nutrition label for formula #1 (*click on the Print Label button, click Print, to return to the label format menu, click on the Close button*).

9. Return to the Nutrition Info & Labeling section of the program (*click on the Return button*).

10. Enter the percentage of each ingredient for formula #2 in the % (wt/wt) column.

11. Enter a name and save formula #2 (*Click on the Formula Name window, enter "food analysis formula #2" in the top Formula Name window, click on the X to close the window, select Save Formula from the File menu*).

12. View and print the nutrition label for formula #2 (*click on the View Label button, click on the Print Label button, then click on Print; to return to the label format menu click on the Close button*).

METHOD B: ADDING NEW INGREDIENTS TO A FORMULA AND DETERMINING HOW THEY INFLUENCE THE NUTRITION LABEL

Sometimes it may be necessary to add additional ingredients to a formula. As an example, let us say you decided to add an additional source of calcium to yogurt formula #1. After contacting several suppliers you decided to add Fieldgate Natural Dairy Calcium 1000, a calcium phosphate product produced by First District Association (Litchfield, MN), to the yogurt formula. This product is a natural dairy-based whey mineral concentrate and contains 25% calcium. You want to determine how much Dairy Calcium 1000 you need to add to have 50% and 100% of the Daily Value (DV) of calcium in one serving of your yogurt. The composition of the Dairy Calcium 1000 you will add is shown in Table 1-2.

Procedure

1. Add the new ingredient to the database. (*From the Formula Dev menu select View Form Dev Section then select Formula Development, from the File menu select Ingredient Files and then select Edit Current Ingredient File.*)

2. Enter the name of the new ingredient and its composition. (*From the Edit Ingredients menu select Add Ingredient, enter the ingredient name—"Dairy Calcium 1000," click OK, enter the amount of each nutrient in the appropriate column, click on the Finish Edit button.*)

3. Edit the ingredient declaration (what will appear on the ingredient list) for the new ingredient.

1-2 table	Composition of Fieldgate Natural Dairy Calcium 1000 (First District Association)
Component	**Amount**
Ash	75%
Calcium	25,000 mg/100g
Calories	40 cal/100g
Lactose	10%
Phosphorus	13,000 mg/100 g
Protein	4.0%
Sugars	10 g/100 g
Total Carbohydrate	10 g/100 g
Total Solids	92%
Water	8.0%

(*From the Edit Ingredients menu select Edit/View Reference #, Source Information and Declaration, enter the ingredient declaration "WHEY MINERAL CONCENTRATE" in the bottom Ingredient Declaration window, click on the Update Information button, click OK, click on the Close button.*)

4. Save the changes to the ingredient file. (*From the File menu Select Save Ingredient File, return to the Formula Development section by selecting Close Ingredient File from the File menu.*)

5. Open food analysis formula #1 in the Formula Development section of the program. (*From the File menu select Open Formula and select food analysis formula #1, click on the Open button, click on Yes for each question.*)

6. Add the new Dairy Calcium 1000 ingredient to "food analysis formula #1." (*Click on the Add Ingredients button, then select Dairy Calcium 1000 from the ingredient list, click on the Add button, click on the X to close the window.*)

7. Calculate the amount of calcium (mg/100 g) required to meet 50% and 100% of the DV (see example below).

Calcium required

$$= (\text{DV for calcium}/\text{serving size})$$
$$\times\ 100\ g \times \%\ \text{of DV required}$$

Calcium required for 50% of the DV

$$= (1000\ \text{mg}/227\ \text{g}) \times 100\ \text{g} \times 0.50$$
$$= 220\ \text{mg}/100\ \text{g}$$

8. Enter the amount of calcium required in the formula and restrict all ingredients in the formula except skim milk and Dairy Calcium 1000. (*Find calcium in the Properties column and enter 220 in the Minimum and Maximum columns for calcium*—this lets the program know that you want to have 220 mg of calcium per 100 g. *In both the Min and Max columns of the formula ingredients enter 38.201 for milk (3.7% fat), 12.888 for condensed skim milk (35% TS), 11.905 for sweetener, sugar liquid, 0.800 for modified starch, and 0.500 for stabilizer, gelatin*—this lets the program adjust the amount of skim milk and Dairy Calcium 1000 (calcium phosphate) and keeps the amount of all the other ingredients constant. *Click on the Formulate button, click OK.*)

9. Enter a name and save the modified formula (*Click on the Formula Name window, enter "food analysis formula # 1added calcium" in the top Formula Name window, click on the X to close the window, select Save Formula from the File menu*).

10. Transfer the new formula (with added calcium phosphate) to the Nutrition Label section of the program. (*From the Formula Dev menu select Send Formula Info to Nutr Label Section, click Yes for each of the questions.*)

11. Make sure your have the correct serving size information. (*See Method A, step 4.*)

12. View and print the nutrition label for the new formula for 50% of the calcium DV. (*Click on the View Label button, click on the Print Label button, then click on Print; click on the Return button to return to the Formula Development Section.*)

13. Produce a formula and label that has 100% of the calcium DV. (*Repeat Steps 7–12 except using the calculated amount of calcium required to meet 100% of the calcium DV*—you will have to perform this calculation yourself by following the example in Step 7.)

METHOD C: AN EXAMPLE OF REVERSE ENGINEERING IN PRODUCT DEVELOPMENT

Procedure

In this example the program will automatically go through the reverse engineering process. Start the example by selecting Cultured Products Automated Examples from the Help menu and clicking on example #4. During this example you proceed to the next step by clicking on the Next button.

1. The information from the nutrition label for the product you want to reverse engineer is entered into the program. (*Comments: In this example serving size, calories, calories from fat, total fat, saturated fat, cholesterol, sodium, total carbohydrate, sugars, protein, vitamin A, vitamin C, calcium, and iron are entered.*)

2. The minimum and maximum levels of each nutrient are calculated on a 100-g basis. (*Comments: The program uses the rounding rules to determine the possible range of each nutrient on a 100-g basis.*)

3. The information about nutrient minimum and maximums is transferred into the Formula Development section of the program. (*Comments: The program has now converted nutrient range information into a form it can use during the formulation process.*)

4. Ingredients used in the formula are then selected based on the ingredient declaration statement on the nutrition label. (*Comments: Selecting the right ingredients can be difficult and an extensive understanding of the ingredient declaration rules is necessary. Additionally some of the required ingredients may not be in the database and will need to be added.*)

5. Restrictions on the amount of each ingredient in the formula are imposed whenever possible.

(Comments: This is a critical step that requires knowledge about the typical levels of ingredients used in the product. Additionally, based on the order of ingredients in the ingredient declaration, approximate ranges can be determined. In this example the amount of modified starch is limited to 0.80%, the amount of gelatin is limited to 0.50%, and the amount of culture is limited to 0.002%.)

6. The program calculates an approximate formula. *(Comments: The program uses the information on nutrient ranges and composition of the ingredients to calculate the amount of each ingredient in the formula.)*

7. The program compares the nutrition information for the developed formula to the original nutrition label. *(Comments: This information is viewed in the Nutrition Label to Formula Spec section of the program accessed by selecting View Reverse Engineering Section then Label to Spec from the Reverse Engineering menu.)*

Questions

1. Based on the labels you produced for yogurt formula #1 and formula #2 in Method A, what nutrient content claims could you make for each formula (a description of nutrient content claims is found in Tables 3–7 and 3–8 in the Nielsen *Food Analysis* text)?

2. How much Dairy Calcium 1000 did you have to add to the yogurt formula to have 50 and 100% of the DV of calcium in the formula?

3. If Dairy Calcium 1000 costs $ 2.50/lb and you are going to have 100% of the DV for calcium in your yogurt, how much extra will you have to charge for a serving of yogurt to cover the cost of this ingredient?

4. Assume you added enough Dairy Calcium 1000 to claim 100% of the DV of calcium, would you expect the added calcium to cause any texture changes in the yogurt?

5. Make a nutrition label for your favorite chocolate chip cookie recipe. You will have to gather several pieces of information including: the weight of each ingredient (not cups, tsp, etc.), the composition of any ingredient that is not in the database, and moisture loss that occurs during baking.

REFERENCES

Nielsen, S.S., and L.E. Metzger. 2003. Nutrition labeling. Ch. 3, in *Food Analysis*, 3rd ed. S.S. Nielsen (Ed.), Kluwer Academic, New York.

Owl Software, 2002. TechWizard™ Version 3 Manual, LaVic, Inc., Lancaster, PA.

Assessment of Accuracy and Precision

INTRODUCTION

Background

Volumetric glassware, mechanical pipettes, and balances are used in many analytical laboratories. If the basic skills are mastered in the use of this glassware and equipment, laboratory exercises are easier, more enjoyable, and the results obtained are more accurate and precise. Measures of accuracy and precision can be calculated based on data generated from the use of glassware and equipment, to evaluate both the skill of the user and how well one can rely on the instrument or glassware.

Determining mass on an analytical balance is the most basic measurement made in an analytical laboratory. Determining and comparing mass is fundamental to assays such as moisture and fat determination. Accurately weighing reagents is the first step in preparing solutions for use in various assays.

Accuracy and precision of the analytical balance is better than any other instrument commonly used to make analytical measurements, provided the balance is properly calibrated and the laboratory personnel use proper technique. With proper calibration and technique, accuracy and precision are limited only by the readability of the balance. Repeatedly weighing a standard weight can yield valuable information about the calibration of the balance and the technician's technique.

Once the performance of the analytical balance and the technician using it have proven acceptable, determination of mass can be used to assess the accuracy and precision of other analytical instruments. All analytical laboratories use volumetric glassware and mechanical pipettes. Mastering their use is necessary to obtain reliable analytical results. To report analytical results from the laboratory in a scientifically justifiable manner, it is necessary to understand accuracy and precision.

A procedure or measurement technique is validated by generating numbers that estimate their accuracy and precision. This laboratory includes assessing the accuracy and precision of automatic pipettors. An example application is determining the accuracy of automatic pipettors in a research or quality assurance laboratory, to help assess their reliability and determine if repair of the pipettors is necessary. Laboratory personnel should periodically check the pipettors to determine if they accurately dispense the intended volume of water. To do this, water dispensed by the pipettor is weighed, and the weight is converted to a volume measurement using the appropriate density of water based on the temperature of the water. If replicated volume data indicate a problem with the accuracy and/or precision of the pipettor, repair is necessary before the pipettor can be used again reliably.

It is generally required that reported values minimally include the mean, a measure of precision, and the number of replicates. The number of significant figures used to report the mean reflects the inherent uncertainty of the value, and it needs to be justified based on the largest uncertainty in making the measurements of relative precision of the assay. The mean value often is expressed as part of a confidence interval (CI) to indicate the range within which the true mean is expected to be found. Comparison of the mean value or the CI to a standard or true value is the first approximation of accuracy. A procedure or instrument is generally not deemed inaccurate if the CI overlaps the standard value. Additionally, a CI that is considerably greater than the readability indicates that the technician's technique needs improvement. In the case of testing the accuracy of an analytical balance with a standard weight, if the CI does not include the standard weight value, it would suggest that either the balance needs calibration or that the standard weight is not as originally issued. Accuracy is sometimes estimated by the relative error ($\%E_{rel}$) between the mean analysis value and the true value. However, $\%E_{rel}$ only reflects tendencies, and in practice often is calculated even when there is no statistical justification that the mean and true value differ. Also, note that there is no consideration of number of replicates in the calculation of $\%E_{rel}$, suggesting that the number of replicates will not affect this estimation of accuracy to any large extent. Absolute precision is reflected by the standard deviation, while relative precision is calculated as the coefficient of variation (CV). Calculations of precision are largely independent of the number of replicates, except that more replicates may give a better estimate of the population variance.

Validation of a procedure or measurement technique can be performed, at the most basic level, as a single trial validation, as is described in this laboratory that includes estimating the accuracy and precision of commonly used laboratory equipment. However, for more general acceptance of procedures, they are validated by collaborative studies involving several laboratories. Collaborative evaluations are sanctioned by groups such as AOAC International, The American Association of Cereal Chemists (AACC) and the American Oil Chemists Society (AOCS). Such collaborative studies are prerequisite to procedures appearing as approved methods in manuals published by these organizations.

Reading Assignment

Literature on how to properly use balances, volumetric glassware, and mechanical pipettes.

Smith, J.S. 2003. Evaluation of analytical data. Ch. 4, in *Food Analysis*, 3rd ed. S.S. Nielsen (Ed.), Kluwer Academic, New York.

Objective

Familiarize, or refamiliarize, oneself with the use of balances, mechanical pipettes, and volumetric glassware, and assess accuracy and precision of data generated.

Principle of Method

Proper use of equipment and glassware in analytical tests helps ensure more accurate and precise results.

Supplies

- 3 Beakers, one each of 20 or 30 ml, 100 ml, and 250 ml
- Buret, 25 or 50 ml
- Erlenmeyer flask, 500 ml
- Funnel, approx. 2 cm diameter (to fill buret)
- Mechanical pipettor, 1000 μl, with plastic tips
- Plastic gloves
- Ring stand and clamps (to hold buret)
- Rubber bulb or pipette pull-up
- Standard weight, 50 or 100 g
- Thermometer, to read near room temperature
- Volumetric flask, 100 ml
- 2 Volumetric pipettes, one each of 1 and 10 ml

Equipment

- Analytical balance
- Top loading balance

Notes

Before or during the laboratory exercise, the instructor is encouraged to discuss the following: (1) difference between dispensing from a volumetric pipette and a graduated pipette, (2) difference between markings on a 10-ml versus a 25- or 50-ml buret.

PROCEDURES

(Record data in tables that follow.)

1. Obtain ~400 ml deionized distilled (dd) H_2O in a 500-ml Erlenmeyer flask for use during this laboratory session. Check the temperature of the water with a thermometer.
2. Analytical balance and volumetric pipettes.
   a. Tare a 100-ml beaker, deliver 10 ml of water from a volumetric pipette into the beaker, and record the weight. Repeat this procedure of taring the beaker, adding 10 ml, and recording the weight, to get six determinations on the same pipette. (Note that the total volume will be 60 ml.) (It is not necessary to empty the beaker after each pipetting.)
   b. Repeat the procedure as outlined in Step 2a but use a 20- or 30-ml beaker and a 1.0-ml volumetric pipette. Do six determinations.
3. Analytical balance and buret.
   a. Repeat the procedure as outlined in Step 2a, but use a 100-ml beaker, a 50-ml (or 25-ml) buret filled with water, and dispense 10 ml of water (i.e., tare a 100-ml beaker, deliver 10 ml of water from the buret into the beaker, and record the weight). (Handle the beaker wearing gloves, to keep oils from your hands off the beaker.) Repeat this procedure of taring the beaker, adding 10 ml, and recording the weight, to get six determinations on the buret. (Note that the total volume will be 60 ml.) (It is not necessary to empty the beaker after each addition.)
   b. Repeat the procedure as outlined in Step 3a but use a 20- or 30-ml beaker and a 1.0-ml volume from the buret. Do six determinations.
4. Analytical balance and mechanical pipette. Repeat the procedure as outlined in Step 2a but use a 20- or 30-ml beaker and a 1.0-ml mechanical pipette (i.e., tare a 20- or 30-ml beaker, deliver 1 ml of water from a mechanical pipettor into the beaker, and record the weight). Repeat this procedure of taring the beaker, adding 1 ml, and recording the weight to get six determinations on the same pipettor. (Note that the total volume will be 6 ml.) (It is not necessary to empty the beaker after each pipetting.)
5. Total content (TC) versus total delivery (TD). Tare a 100-ml volumetric flask on a top loading balance. Fill the flask to the mark with water. Weigh the water in the flask. Now tare a 250-ml beaker and pour the water from the volumetric into the beaker. Weigh the water delivered from the volumetric flask.
6. Readability versus accuracy. Zero a top loading balance and weigh a 100-g (or 50-g) standard weight. Record the observed weight. Use gloves or finger cots as you handle the standard weight to keep oils from your hands off the weight. Repeat with the same standard weight on at least two other top loading balances, recording the observed weight and the type and model (e.g., Mettler, Sartorius) of balance used.

DATA AND CALCULATIONS

Calculate the exact volume delivered in Parts 2–5, using each weight measurement and the known density of

2-1 table	Viscosity and Density of Water at Various Temperatures				
Temp. (°C)	Density (g/ml)	Viscosity (cps)	Temp. (°C)	Density (g/ml)	Viscosity (cps)
20	0.99823	1.002	24	0.99733	0.9111
21	0.99802	0.9779	25	0.99707	0.8904
22	0.99780	0.9548	26	0.99681	0.8705
23	0.99757	0.9325	27	0.99654	0.8513

water (see Table 2-1). Using *volume* data, calculate the following indicators of accuracy and precision: mean, standard deviation, coefficient of variation, percent relative error, 95% confidence interval. Use your first three measurements for $n = 3$ values requested, and all six measurements for $n = 6$ values.

Data for Parts 2, 3, and 4:

	Volumetric Pipette				Buret				Mechanical Pipettor	
	1 ml		10 ml		1 ml		10 ml		1 ml	
Trial	Wt.	Vol.	Wt.	Vol.	Wt.	Vol.	Wt.	Vol.	Wt.	Vol.
1										
2										
3										
4										
5										
6										
$n = 3$										
Mean	—		—		—		—		—	
S	—		—		—		—		—	
CV	—		—		—		—		—	
%E_{rel}	—		—		—		—		—	
$CI_{95\%}$	—		—		—		—		—	
$n = 6$										
Mean	—		—		—		—		—	
S	—		—		—		—		—	
CV	—		—		—		—		—	
%E_{rel}	—		—		—		—		—	
$CI_{95\%}$	—		—		—		—		—	

Part 5 data:

	Wt.	Vol.
Water in flask =		
Water in beaker =		

Part 6 data:

Balance	Type/Model of Balance	Wt. of Standard Weight
1		
2		
3		

Questions

1. Theoretically, how are standard deviation, coefficient of variation, mean, percent relative error, and 95% confidence interval affected by: (1) more replicates, and (2) a larger size of the measurement? Was this evident in looking at the actual results obtained using the volumetic pipettes and the buret, with $n = 3$ versus $n = 6$, and with 1 ml versus 10 ml? (see table below)

2. Why are percent relative error and coefficient of variation used to compare the accuracy and precision, respectively, of the volumes from pipetting/dispensing 1 and 10 ml with the volumetric pipettes and buret in Parts 2 and 3, rather than simply the mean and standard deviation, respectively?

3. Compare and discuss the accuracy and the precision of the volumes from the 1 ml pipetted/dispensed using a volumetric pipette, buret, and mechanical pipettor (Parts 2, 3, and 4). Are these results consistent with what would be expected?

4. If accuracy and/or precision using the mechanical pipettor are less than should be expected, what could you do to improve its accuracy and precision?

5. In a titration experiment using a buret, would you expect to use much less than a 10-ml volume in each titration? Would you expect your accuracy and precision to be better using a 10-ml buret or a 50-ml buret? Why?

	Theoretical		Actual, with Results Obtained	
	More Replicates	Larger Measurement	More Replicates	Larger Measurement
Standard deviation				
Coefficient of variation				
Mean				
Percent relative error				
95% Confidence interval				

6. How do your results from Part #5 of this lab differentiate "to contain" from "to deliver"? Is a volumetric flask "to content" or "to deliver"? Which is a volumetric pipette?
7. From your results from Part #6 of this lab, would you now assume that since a balance reads to 0.01 g that it is accurate to 0.01 g?
8. What sources of error (human and instrumental) were evident or possible in Parts #2–4, and how could these be reduced or eliminated? Explain.
9. You are considering adopting a new analytical method in your lab to measure the moisture content of cereal products. How would you determine the precision of the new method and compare it to the old method? How would you determine (or estimate) the accuracy of the new method?

ACKNOWLEDGMENT

This laboratory was developed with input from Dr Charles E. Carpenter, Department of Nutrition and Food Sciences, Utah State University, Logan, UT.

REFERENCES

Smith, J.S. 2003. Evaluation of analytical data. Ch. 4, in *Food Analysis*, 3rd ed. S.S. Nielsen (Ed.). Kluwer Academic, New York.

3
chapter

Determination of Moisture Content

INTRODUCTION

Background

The moisture (or total solids) content of foods is important to food manufacturers for a variety of reasons. Moisture is an important factor in food quality, preservation, and resistance to deterioration. Determination of moisture content also is necessary to calculate the content of other food constituents on a uniform basis (i.e., dry weight basis). The dry matter that remains after moisture analysis is commonly referred to as total solids.

While moisture content is not given on a nutrition label, it must be determined to calculate total carbohydrate content. Moisture content of foods can be determined by a variety of methods, but obtaining accurate and precise data is commonly a challenge. The carious method of analysis have different applications, advantages, and disadvantages (see Reading Assignment). If the ash content also is to be determined, it is often convenient to combine the moisture and ash determinations. In this experiment, several methods to determine the moisture content of foods will be used and the results compared. Summarized below are the food samples proposed for analysis and the methods used. However, note that other types of food samples could be analyzed and groups of students could analyze different types of food samples. It is recommended that all analyses be performed in triplicate, as time permits.

	Corn Syrup	Corn Flour	Milk (liquid)	Nonfat Dry Milk	Basil
Forced draft oven	X	X	X	X	X
Vacuum oven	X				
Microwave drying	X		X		
Rapid moisture analyzer		X	X		
Toluene distillation				X	X
Karl Fischer		X	X	X	
Near infrared		X			

Reading Assignment

Bradley, R.L., Jr. 2003. Moisture and total solids analysis, Ch. 6, in *Food Analysis*, 3rd ed. S.S. Nielsen (Ed.), Kluwer Academic, New York.

Overall Objective

The objective of this experiment is to determine and compare the moisture contents of foods by various methods of analysis.

METHOD A: FORCED DRAFT OVEN

Objective

Determine the moisture content of corn syrup and corn flour using a forced draft oven method.

Principle of Method

The sample is heated under specified conditions and the loss of weight is used to calculate the moisture content of the sample.

Supplies

- Basil (fresh), 15 g (ground)
- Beaker, 25–50 ml (to pour corn syrup into pans)
- Corn flour, 10 g
- Corn syrup, 15 g
- 3 Crucibles (preheated at 550°C for 24 hr)
- 2 Desiccators (with dried desiccant)
- Liquid milk, 20 ml
- Nonfat dry milk (NFDM), 10 g
- Plastic gloves (or tongs)
- 2 Spatulas
- 5 Trays (to hold/transfer samples)
- 2 Volumetric pipettes, 5 ml
- 6 Weighing pans—disposable aluminum open pans (for use with corn syrup) (pre-dried at 100°C for 24 hr)
- 6 Weighing pans—metal pans with lids (for use with corn flour and NFDM) (pre-dried at 100°C for 24 hr)

Equipment

- Forced draft oven
- Analytical balance, 0.1 mg sensitivity

Note

Glass microfiber filters (e.g., GF/A, Whatman, Newton, MA), predried for 1 hr at 100°C, can be used to cover samples to prevent splattering in the forced draft oven and the vacuum oven. Instructors may want to have students compare results with and without these fiberglass covers.

Procedure

Instructions are given for analysis in triplicate.

I. Moisture in Corn Syrup

1. Label dried pans (disposable aluminum open pans) and weigh accurately.
2. Place 5 g of sample in the pan and weigh accurately.
3. Place in a forced draft oven at 98–100°C for 24 hr.

4. Store in a desiccator until samples are weighed.
5. Calculate percentage moisture (wt/wt) as described below.

II. Moisture in Corn Flour (Method 44-15A of American Association of Cereal Chemists, one-stage procedure)

1. Weigh accurately dried pan with lid. (Note identifier number on pan and lid.)
2. Place 2–3 g of sample in the pan and weigh accurately.
3. Place in a forced draft oven at 130°C for 1 hr. Be sure metal covers are ajar, to allow water loss.
4. Remove from oven, realign covers to close, cool, and store in desiccator until samples are weighed.
5. Calculate percentage moisture (wt/wt) as described below.

III. Moisture in Liquid Milk (AOAC Method 990.19, 990.20)

1. Label and weigh accurately pre-dried crucibles (550°C for 24 hr). (Note identified number on crucible.)
2. Place 5 g of sample in the crucible and weigh accurately.
3. Evaporate a majority of water on a hot plate; do not dry the sample completely.
4. Place in a forced draft oven at 100°C for 3 hr.
5. Store in a desiccator until samples are weighed.
6. Calculate percentage moisture (wt/wt) as described below.

Note: Ash content of this milk sample could be determined by placing the milk sample, dried at 100°C for 3 hr., in a muffle furnace at 550°C for 18–24 hr. After cooling in a desiccator, the crucibles containing ashed milk would be weighed and the ash content calculated.

IV. Moisture of Nonfat Dry Milk

1. Weigh accurately the dried pan with lid. (Note identifier number on pan and lid.)
2. Place 3 g of sample in the pan and weigh accurately.
3. Place pan in a forced draft oven at 100°C for 24 hr.
4. Store in a desiccator until samples are weighed.
5. Calculate percentage moisture (wt/wt) as described below.

V. Moisture in Fresh Basil

1. Label dried pans (disposable aluminum open pans) and weigh accurately.

2. Place 3 g of ground sample in the pan and weigh accurately.
3. Place in a forced draft oven at 98–100°C for 24 hr.
4. Store in a desiccator until samples are weighed.
5. Calculate percentage moisture (wt/wt) as described below.

Data and Calculations

Calculate percentage moisture (wt/wt):

$$\% \text{ moisture} = \frac{\text{wt of H}_2\text{O in sample}}{\text{wt of wet sample}} \times 100$$

$$\% \text{ moisture}$$
$$= \frac{\left(\begin{array}{l}\text{wt of wet sample}\\ + \text{pan}\end{array}\right) - \left(\begin{array}{l}\text{wt of dried sample}\\ + \text{pan}\end{array}\right)}{(\text{wt of wet sample} + \text{pan}) - (\text{wt of pan})} \times 100$$

$$\% \text{ ash, wet weight basis (wwb)} = \frac{\text{wt of ash}}{\text{wt of wet sample}} \times 100$$

$$\% \text{ ash, wwb}$$
$$= \frac{\left(\begin{array}{l}\text{wt of dry sample}\\ + \text{crucible}\end{array}\right) - \left(\begin{array}{l}\text{wt of ashed sample}\\ + \text{crucible}\end{array}\right)}{(\text{wt of wet sample} + \text{crucible}) - (\text{wt of crucible})} \times 100$$

Sample	Trial	Pan (g)	Pan + Wet Sample (g)	Pan + Dried Sample (g)	% Moisture
Corn syrup	1				
	2				
	3				
					$\overline{X} =$
					SD $=$
Corn flour	1				
	2				
	3				
					$\overline{X} =$
					SD $=$
Liquid milk	1				
	2				
	3				
					$\overline{X} =$
					SD $=$
Nonfat dry milk	1				
	2				
	3				
					$\overline{X} =$
					SD $=$
Fresh basil	1				
	2				
	3				
					$\overline{X} =$
					SD $=$

METHOD B: VACUUM OVEN

Objective

Determine the moisture content of corn syrup by the vacuum oven method, with and without the addition of sand to the sample.

Principle

The sample is heated under conditions of reduced pressure to remove water and the loss of weight is used to calculate the moisture content of the sample.

Supplies

- Corn syrup, 30 g
- Desiccator (with dried desiccant)
- 3 Glass stirring rods (ca. 2–3 cm long, pre-dried at 100°C for 3 hr)
- Plastic gloves (or tongs)
- Pipette bulb or pump
- Sand, 30 g (pre-dried at 100°C for 24 hr)
- 2 Spatulas
- Volumetric pipette, 5 ml
- 6 Weighing pans—disposable aluminum open pans (pre-dried at 100°C for 3 hr)

Equipment

- Vacuum oven (capable of pulling vacuum to <100 mm of mercury)
- Analytical balance, 0.1 mg sensitivity

Procedure

I. Moisture of Corn Syrup, without Use of Drying Sand

1. Label weighing pans (i.e., etch identifier into tab of disposable aluminum pan) and weigh accurately.
2. Place 5 g of sample in the weighing pan and weigh accurately.
3. Dry at 70°C and a vacuum of at least 26 in. for 24 hrs, but pull and release the vacuum slowly. (Note that samples without drying sand will bubble up and mix with adjoining samples if pans are too close together.) Bleed dried air into the oven as vacuum is released.
4. Store in a desiccator until samples are cooled to ambient temperature. Weigh.

II. Moisture of Corn Syrup, with Use of Drying Sand

1. Label weighing pan, add 10 g dried sand and stirring rod, then weigh accurately.
2. Add 5 g of sample and weigh accurately. Add 5 ml of deionized distilled (dd) water. Mix with stirring rod being careful not to spill any of the sample. Leave the stirring bar in the pan.
3. Dry at 70°C and a vacuum of <100 mm mercury for 24 hr. Bleed dried air into the oven as vacuum is released.
4. Store in a desiccator until samples are cooled to ambient temperature. Weigh.

Data and Calculations

Calculate percentage moisture (wt/wt) as in Method A.

Sample	Trial	Pan (g)	Pan + Wet Sample (g)	Pan + Dried Sample (g)	% Moisture
Corn syrup without sand	1				
	2				
	3				$\overline{X} =$ SD =
Corn syrup with sand	1				
	2				
	3				$\overline{X} =$ SD =

METHOD C: MICROWAVE DRYING OVEN

Objective

Determine the moisture content of corn syrup and milk (liquid) using a microwave drying oven.

Principle

The sample is heated using microwave energy, and the loss of weight is used to calculate the moisture content of the sample.

Supplies

- Corn syrup, 4 g
- Glass stirring rod (to spread corn syrup)
- Milk (liquid), 4 g
- 6 Paper pads (for use in microwave oven)
- Pasteur pipette and bulb (to spread milk sample)
- Plastic gloves

Equipment

- Microwave drying oven (e.g., from CEM Corporation, Matthew, NC).

Procedure

Follow instructions from manufacturer for use of the microwave drying oven, regarding the following:

- Turning on instrument and warming up
- Loading method for specific application (i.e., sets time, power, etc.)
- Taring instrument
- Testing sample
- Obtaining results

Data and Calculations

Sample	Trial	% Moisture	g Water/ g Dry Matter
Corn syrup	1		
	2		
	3		
		$\overline{X} =$	$\overline{X} =$
		SD =	SD =
Milk (liquid)	1		
	2		
	3		
		$\overline{X} =$	$\overline{X} =$
		SD =	SD =

METHOD D: RAPID MOISTURE ANALYZER

Objective

Determine the moisture content of corn flour using a rapid moisture analyzer.

Principle

The sample sitting on a digital balance is heated under controlled high heat conditions, and the instrument automatically measures the loss of weight to calculate the percentage moisture or solids.

Supplies

- Corn flour, 10 g
- Milk, 10 ml
- Plastic gloves
- Spatula

Equipment

- Rapid Moisture Analyzer (e.g., from Compu-trac, Arizona Instrument Corp., Tempe, AZ)

Procedure

Follow instructions from manufacturer for use of the rapid moisture analyzer, regarding the following:

- Turning on instrument and warming up
- Select test material
- Taring instrument
- Testing sample
- Obtaining results

Data and Calculations

Sample	% Moisture 1	2	3	Mean
Corn flour				
Milk				

METHOD E: TOLUENE DISTILLATION

Objective

Determine the moisture content of basil by the toluene distillation method.

Principle

The moisture in the sample is codistilled with toluene, which is immiscible in water. The mixture that distills off is collected, and the volume of water removed is measured.

Chemicals

	CAS No.	Hazards
Toluene	108-88-3	Harmful, highly flammable

Hazards, Cautions, and Waste Disposal

Toluene is highly flammable and is harmful if inhaled. Use adequate ventilation. Wear safety glasses and gloves at all times. For disposal of toluene waste, follow good laboratory practices outlined by environmental health and safety protocols at your institution.

Supplies

- Fresh basil, 40–50 g
- NFDM, 40–50 g
- Toluene, A.C.S. grade

Equipment

- Analytical balance, 0.1 mg sensitivity

- Glass distillation apparatus with ground glass joints: (1) Boiling flask, 250 or 300 ml, round-bottom, shortneck flask with a T.S.24/40 joint; (2) West condenser with drip tip, 400 mm in length with a T.S. 24/40 joint; (3) Bidwell–Sterling trap, T.S. 24/40 joint, 3 ml capacity graduated in 0.1 ml intervals
- Heat source, capable of refluxing toluene in the apparatus above (e.g., heating mantle connected to voltage controller). No open flame!
- Nylon bristle buret brush, $\frac{1}{2}$ in. in diameter, and a wire loop. (It should be long enough to extend through the condenser, ca. 450 mm. Flatten the loop on the buret brush and use this brush, inverted, as a wire to dislodge moisture drops in the moisture trap.)

Procedure

1. Grind the fresh basil with a small table-top food grinder. Pulse grind the sample in 5–10 s intervals. Avoid long pulses and excessive grinding to prevent frictional heat.
2. Weigh approximately 40 g of sample (basil or NFDM) accurately (amount chosen to yield 2–5 ml water)
3. Transfer sample quantitatively to distilling flask. Add sufficient toluene to cover the sample completely (not less than 75 ml).
4. Assemble the apparatus as shown in Chapter 6 of Nielsen's textbook. Fill the trap with toluene by pouring it through the condenser until it just fills the trap and begins to flow into the flask. Insert a loose nonabsorbing cotton plug into the top of the condenser to prevent condensation of atmospheric moisture in the condenser.
5. Bring to boil and reflux at about 2 drops per second until most of the water has been collected in the trap, then increase the reflux rate to ca. 4 drops per second.
6. Continue refluxing until two consecutive readings 15 min apart show no change. Dislodge any water held up in the condenser with a brush or wire loop. Rinse the condenser carefully with ca. 5 ml toluene. Dislodge any moisture droplets adhering to the Bidwell–Sterling trap or toluene trapped under the collected moisture. For this, use the wire. Rinse wire with a small amount (10 ml) of toluene before removing from apparatus.
7. Continue refluxing for 3–5 min, remove the heat, and cool the trap to 20°C in a suitable water bath.
8. Calculate the moisture content of the sample:

% moisture

$$= [\text{vol. of water (ml)}/\text{wt of sample (g)}] \times 100$$

Notes

1. Flask, condenser, and receiver must be scrupulously clean and dry. For example, the apparatus, including the condenser, could be cleaned with potassium dichromate-sulfuric acid cleaning solution, rinsed with water, rinsed with 0.05 N potassium hydroxide solution, rinsed with alcohol, then allowed to drain for 10 min. This procedure will minimize the adherence of water droplets to the surfaces of the condenser and the Bidwell–Sterling trap.
2. A correction blank for toluene must be conducted periodically by adding 2–3 ml of distilled water to 100 ml of toluene in the distillation flask, then following the procedure in steps 2–6 above.

Data and Calculations

Wt. Sample (g)	Vol. Water Removed (ml)	% Moisture

METHOD F: KARL FISCHER

Objective

Determine the moisture content of NFDM and corn flour by the Karl Fischer (KF) method.

Principle

When the sample is titrated with the KF reagent, which contains iodine and sulfur dioxide, the iodine is reduced by sulfur dioxide in the presence of water from the sample. The water reacts stoichiometrically with the KF reagent. The volume of KF reagent required to reach the endpoint of the titration (visual, conductometric, or coulometric) is directly related to the amount of water in the sample.

Chemicals

	CAS No.	Hazards
Karl Fischer reagent		Toxic
2-methoxyethanol	109-86-4	
Pyridine	110-86-1	
Sulfur dioxide	7446-09-5	
Iodine	7553-56-2	Harmful, dangerous to the environment
Methanol, anhydrous	67-56-1	Extremely flammable
Sodium tartrate dihydrate ($Na_2C_4O_6 \cdot 2H_2O$)	868-18-8	

Reagents

- KF reagent
- Methanol, anhydrous
- Sodium tartrate dihydrate, 1 g, dried at 150°C for 2 hr

Hazards, Cautions, and Waste Disposal

Use the anhydrous methanol in an operating hood, since the vapors are harmful and it is toxic. Otherwise, adhere to normal laboratory safety procedures. Use appropriate eye and skin protection. The KF reagent and anhydrous methanol should be disposed of as hazardous wastes.

Supplies

- Corn flour
- Graduated cylinder, 50 ml
- NFDM
- 2 Spatulas
- Weighing paper

Equipment

- Analytical balance, 0.1 g sensitivity
- Drying oven
- KF titration unit (e.g., from Barnsted Themaline, Berkeley, CA, Aquametry Apparatus)

Procedure

Instructions are given as for a nonautomated unit, and for analysis in triplicate.

I. Apparatus Set Up

Assemble titration apparatus and follow instructions of manufacturer. The titration apparatus includes the following: buret; reservoir for reagent; magnetic stirring device; reaction/titration vessel; electrodes; and circuitry for dead stop endpoint determination. Note that the reaction/titration vessel of the KF apparatus (and the anhydrous methanol within the vessel) must be changed after analyzing several samples (exact number depends on type of sample). Remember that this entire apparatus is very fragile. To prevent contamination from atmospheric moisture, all openings must be closed and protected with drying tubes.

II. Standardizing Karl Fischer Reagent

The KF reagent is standardized to determine its water equivalence. Normally, this needs to be done only once a day, or when changing the KF reagent supply.

1. Add approximately 50 ml of anhydrous methanol to reaction vessel through the sample port.
2. Put the magnetic stir bar in the vessel and turn on the magnetic stirrer.
3. Remove the caps (if any) from drying tube. Turn the buret stopcock to the *filling position*. Hold one finger on the air-release hold in the rubber bulb and pump the bulb to fill the buret. Close the stopcock when the KF reagent reaches the desired level (at position 0.00 ml) in the buret.
4. Titrate the water in the solvent (anhydrous methanol) by adding enough KF reagent to just change the color of the solution from clear or yellow to dark brown. This is known as the KF endpoint. Note and record the conductance meter reading. (You may titrate to any point in the brown KF zone on the meter, but make sure that you always titrate to that same endpoint for all subsequent samples in the series.) Allow the solution to stabilize at the endpoint on the meter for at least 1 min before proceeding to the next step.
5. Weigh, to the nearest milligram, approximately 0.3 g of sodium tartrate dihydrate, previously dried at 150°C for 2 hr.
6. Fill the buret with the KF reagent, then titrate the water in the sodium tartrate dihydrate sample as in step II.4. Record the volume (ml) of KF reagent used.
7. Calculate the KF reagent water (moisture) equivalence (KFR$_{eq}$) in mg H$_2$O/ml:

$$KFR_{eq} = \frac{36 \text{ g/mol} \times S \times 1{,}000}{230.08 \text{ g/mol} \times A}$$

where:

S = weight of sodium tartrate dihydrte (g)
A = ml of KF reagent required for titration of sodium tartrate dihydrate.

III. Titration of Sample

1. Prepare samples for analysis and place in reaction vessel as described below.

   If samples are in powder form:

   a. Use an analytical balance to weigh out approximately 0.3 g of sample, and record the exact sample weight (S) to the nearest milligram.

b. Remove the conductance meter from the reaction vessel, then transfer your sample to the reaction vessel through the sample port immediately. (Use an extra piece of weighing paper to form a cone-shaped funnel in the sample port, then pour your sample through the funnel into the reaction vessel.)

c. Put the conductance meter and stopper back in the reaction vessel. The color of the solution in the vessel should change to light yellow and the meter will register below the KF zone on the meter.

If any samples analyzed are in liquid form:

a. Use a 1 ml syringe to draw up about 0.1 ml of sample. Weigh the syringe with sample on an analytical balance and record the exact weight (S_1) to the nearest milligram.

b. Inject 1–2 drops of sample into the reaction vessel through the sample port, then weigh the syringe again (S_0), to the nearest milligram.

c. Sample weight (S) is the difference of S_1 and S_0.

$$S = S_1 - S_0$$

d. Put the stopper back in the sample port of the reaction vessel. The color of the solution in the vessel should change to light yellow and the meter will register below the KF zone on the meter.

2. Fill the buret, then titrate the water in the sample as in step II.4 above. Record the volume (ml) of KF reagent used.

3. To titrate another sample, repeat steps II.5–II.7 above with the new sample. After titrating several samples (exact number depends on the nature of the sample), it is necessary to start with fresh methanol in a clean reaction vessel. Record the volume (ml) of KF reagent used for each titration.

Data and Calculations

Calculate the moisture content of the sample as follows:

$$\%H_2O = \frac{KFR_{eq} \times K_s}{S} \times 100$$

where:

KFR$_{eq}$ = Water equilance of KF reagent (mg) H_2O/ml

K_s = ml of KF reagent required for titration of sample

S = weight of sample (mg)

Karl Fischer reagent water equivalence (KFR$_{eq}$):

Trial	Wt. Sodium tartrate dihydrate (g)	Buret Start (ml)	Buret End (ml)	Volume Titrant (ml)	Calculated KFR$_{eq}$
1					
2					
3					$\overline{X} =$

Calculation for KFR$_{eq}$:

Moisture content of samples by Karl Fischer method:

Sample	Trial	Wt. Sample (g)	Buret Start (ml)	Buret End (ml)	Volume Titrant (ml)	% Moisture

METHOD G: NEAR INFRARED ANALYZER

Objective

Determine the moisture content of corn flour using a near infrared analyzer.

Principle

Specific frequencies of infrared radiation are absorbed by the functional groups characteristic of water (i.e., the –OH stretch of the water molecule). The concentration of moisture in the sample is determined by measuring the energy that is reflected or transmitted by the sample, which is inversely proportional to the energy absorbed.

Supplies

- Corn flour
- Pans and sample preparation tools for near infrared analyzer

Equipment

- Near infrared analyzer

Procedure

Follow instructions from manufacturer for use of the near infrared analyzer, regarding the following:

- Turning on instrument and warming up
- Calibrating instrument
- Testing sample
- Obtaining results

Data and Calculations

Corn Flour % Moisture			
1	2	3	Mean

Questions

1. In separate tables, summarize the results from the various methods used to determine the moisture content of each type of food sample analyzed: (a) corn syrup, (b) liquid milk, (c) corn flour, (d) NFDM, and (e) basil. Include in each table the following for each method: (a) Data from individual determinations, (b) mean value, (c) standard deviation, (d) observed appearance, etc. of samples, (e) relative advantages of method, and (f) relative disadvantages of method.

2. Calculate the moisture content of the liquid milk samples as determined by the forced draft oven and microwave drying oven methods in terms of g H_2O/g dry matter and include this in a table of results.

Method	Liquid Milk Moisture Content	
	Mean % Moisture	Mean g Water/ g Dry Matter
Forced draft oven		
Microwave drying oven		

3. Why was the milk sample partially evaporated on a hot plate before being dried in the hot air oven?
4. Why is there a need to cover the sample during oven drying?
5. Explain the source of pigment in the toluene from moisture analysis of basil. Does this represent any moisture? Why or why not?
6. Explain the theory/principles involved in predicting the concentrations of various constituents in a food sample by NIR analysis. Why do we say "predict" and not "measure"? What assumptions are being made?
7. Your quality control lab has been using a hot air oven method to make moisture determinations on various products produced in your plant. You have been asked to evaluate the feasibility of switching to new methods (the specific one would depend on the product) for measuring moisture content.
   a. Describe how you would evaluate the accuracy and precision of any new method.
   b. What common problems or disadvantages with the hot air oven method would you seek to reduce or eliminate using any new method?
   c. You are considering the use of a toluene distillation procedure or KF titration method for some of your products that are very low in moisture. What are the advantages of each of these methods over the hot air oven method in the proposed use? What disadvantages or potential problems might you encounter with the other two methods?

ACKNOWLEDGMENTS

This experiment was developed in part with materials provided by Dr Charles E. Carpenter, Department of Nutrition and Food Sciences, Utah State University, Logan UT, and by Dr Joseph Montecalvo, Jr., Department of Food Science and Nutrition, California Polytechnic State University, San Luis Obispo, CA. Arizona Instrument Corp., Tempe, AZ, is acknowledged for its partial contribution of a Computrac moisture analyzer for use in developing a section of this laboratory exercise.

REFERENCES

AACC. 2000. *Approved Methods of Analysis*, 10th ed. American Association of Cereal Chemists, St. Paul, MN.

AOAC International. 2000. *Official Methods of Analysis*, 17th ed., AOAC International, Gaithersburg, MD.

Bradley, R.L., Jr. 2003. Moisture and Total Solids Analysis, Ch. 6, in *Food Analysis*, 3rd ed. S.S. Nielsen (Ed.), Kluwer Academic, New York.

Wehr, H.M., and Frank, J.F. (Eds.). 2002. *Standard Methods for the Examination of Dairy Products*, 17th ed., American Public Health Association, Washington, DC.

Determination of Fat Content

Laboratory Developed in part by

Dr Charles Carpenter,
Utah State University, Department of Nutrition and Food Sciences, Logan, Utah

INTRODUCTION

Background

The term "lipid" refers to a group of compounds that are sparingly soluble in water, but show variable solubility in a number of organic solvents (e.g., ethyl ether, petroleum ether, acetone, ethanol, methanol, benzene). The lipid content of a food determined by extraction with one solvent may be quite different from the lipid content as determined with another solvent of different polarity. Fat content is determined often by solvent extraction methods (e.g., Soxhlet, Goldfish, Mojonnier), but it also can be determined by non-solvent wet extraction methods (e.g., Babcock, Gerber), and by instrumental methods that rely on the physical and chemical properties of lipids (e.g., infrared, density, x-ray absorption). The method of choice depends on a variety of factors, including the nature of the sample (e.g., dry versus moist), the purpose of the analysis (e.g., official nutrition labeling or rapid quality control), and instrumentation available (e.g., Babcock uses simple glassware and equipment; infrared requires an expensive instrument).

This experiment includes the Soxhlet, Goldfish, Mojonnier, and Babcock methods. If samples analyzed by these methods can be tested by an instrumental method for which equipment is available in your laboratory, data from the analyses can be compared. Snack foods are suggested for analysis and comparison by the Soxhlet and Goldfish methods, and milk by the Mojonnier and Babcock methods. However, other appropriate foods could be substituted, and results compared between methods. Also, the experiment specifies the use of petroleum ether as the solvent for the Soxhlet and Goldfish methods. However, anhydrous ethyl ether could be used for both methods, but appropriate precautions must be taken.

Reading Assignment

Min, D.B., and Boff, J.M. 2003. Crude fat analysis. Ch. 8, in *Food Analysis*, 3rd ed. S.S. Nielsen (Ed.), Kluwer Academic, New York.

Objective

Determine the lipid contents of various snack food by the Soxhlet and Goldfish methods, and determine the lipid content of milk by the Mojonnier and Babcock methods.

METHOD A: SOXHLET METHOD

Principle of Method

Fat is extracted, semicontinuously, with an organic solvent. Solvent is heated and volatilized, then is condensed above the sample. Solvent drips onto the sample and soaks it to extract the fat. At 15–20 min intervals, the solvent is siphoned to the heating flask, to start the process again. Fat content is measured by weight loss of sample or weight of fat removed.

Chemicals

	CAS No.	Hazards
Petroleum ether	8032-32-4	Harmful, highly flammable, dangerous for environment
(or Ethyl ether)	60-29-7	Harmful, extremely flammable

Hazards, Precautions, and Waste Disposal

Petroleum ether and ethyl ether are fire hazards; avoid open flames, breathing vapors, and contact with skin. Ether is extremely flammable, is hygroscopic, and may form explosive peroxides. Otherwise, adhere to normal laboratory safety procedures. Wear gloves and safety glasses at all times. Petroleum ether and ether liquid wastes must be disposed of in designated hazardous waste receptacles.

Supplies

- 3 Aluminum weighing pans, pre-dried in 70°C vacuum oven for 24 hr
- Beaker, 250 ml
- Cellulose extraction thimbles, pre-dried in 70°C vacuum oven for 24 hr
- Desiccator
- Glass boiling beads
- Glasswool, pre-dried in 70°C vacuum oven for 24 hr
- Graduated cylinder, 500 ml
- Mortar and pestle
- Plastic gloves
- Snack foods (need to be fairly dry and able to be ground with a mortar and pestle)
- Spatula
- Tape (to label beaker)
- Tongs
- Weighing pan (to hold 30 g snack food)

Equipment

- Analytical balance
- Soxhlet extractor, with glassware
- Vacuum oven

Procedure

(Instructions are given for analysis in triplicate.)

1. Record the fat content of your snack food product as reported on the package label. Also record serving size so you can calculate g fat/100 g product.
2. Slightly grind ~30 g sample with mortar and pestle (excessive grinding will lead to greater loss of fat in mortar).
3. Wearing plastic gloves, remove 3 pre-dried cellulose extraction thimbles from the desiccator. Label the thimbles on the outside with your initials and a number (use a lead pencil), then weigh accurately on an analytical balance.
4. Place ~2–3 g of sample in the thimble. Reweigh. Place a small plug of dried glass wool in each thimble. Reweigh.
5. Place the three samples in a Soxhlet extractor. Put ~350 ml petroleum ether in the flask, add several glass boiling beads, and extract for 6 hr or longer. Place a 250-ml beaker labeled with your name below your samples on the Soxhlet extraction unit. Samples in thimbles will be placed in the beaker after extraction and before drying.
6. Remove thimbles from the Soxhlet extractor using tongs, air dry overnight in a hood, then dry in a vacuum oven at 70°C, 25 in. mercury, for 24 hr. Cool dried samples in a desiccator then reweigh.
7. Correct for moisture content of product as follows:
   a. Using the remainder of the ground sample and three dried, labeled, and weighed aluminum sample pans, prepare triplicate 2–3 g samples for moisture analysis.
   b. Dry sample at 70°C, 25 in. mercury, for 24 hr in a vacuum oven.
   c. Reweigh after drying, and calculate moisture content of the sample.

Data and Calculations

Using the weights recorded in the tables below, calculate the percent fat (wt/wt) on a wet weight basis as determined by the Soxhlet extraction method. If the fat content of the food you analyzed was given on the label, report this theoretical value.

    Name of Snack Food:
    Label g fat/serving:
    Label serving size (g):
    Label g fat/100 g product:

Data from Soxhlet extraction:

Trial	Thimble (g)	Wet Sample + Thimble (g)	Wet Sample + Thimble + Glass Wool (g)	Wet Sample (g)	Dry Sample + Thimble + Glass Wool (g)
1					
2					
3					

Data from moisture analysis:

Trial	Pan (g)	Pan + Wet Sample (g)	Pan + Dried Sample (g)	% Moisture
1				
2				
3				
			$\overline{X} =$	
			SD $=$	

Calculation of % moisture:

$$\frac{\left(\begin{array}{c}\text{wt of wet sample}\\ +\,\text{pan}\end{array}\right) - \left(\begin{array}{c}\text{wt of dried sample}\\ +\,\text{pan}\end{array}\right)}{(\text{wt of wet sample } +\text{ pan}) - (\text{wt of pan})} \times 100$$

Trial	% Fat + % Water	Calc. % Fat
1		
2		
3		

Calculation of % fat:

% (fat + moisture)

$$= \frac{\left[\left(\begin{array}{c}\text{Initial wt of sample}\\ +\,\text{Thimble}\\ +\,\text{Glass wool}\end{array}\right) - \left(\begin{array}{c}\text{Final wt of sample}\\ +\,\text{Thimble}\\ +\,\text{Glass wool}\end{array}\right)\right] \times 100}{\text{Initial wt of sample}}$$

% fat (wt/wt) = (%fat + moisture) − (% moisture)

(*Note:* Use average % moisture in this calculation)

Questions

1. The Soxhlet extraction procedure utilized petroleum ether. What were the advantages of using it rather than ethyl ether?
2. What were the advantages of using the Soxhlet extraction method rather than the Goldfish extraction method?
3. If the fat content measured here differed from that reported on the nutrition label, how might this be explained?

METHOD B: GOLDFISH METHOD

Principle

Fat is extracted, continuously, with an organic solvent. Solvent is heated and volatilized, then is condensed above the sample. Solvent continuously drips through the sample to extract the fat. Fat content is measured by weight loss of sample or weight of fat removed.

Chemicals

Same as for Method A, Soxhlet.

Hazards, Precautions, and Waste Disposal

Same as for Method A, Soxhlet.

Supplies

Same as for Method A, Soxhlet.

Equipment

- Goldfish extraction apparatus
- Analytical balance
- Vacuum oven

Procedure

(Instructions are given for analysis in triplicate.)
Note: Analyze samples in triplicate.

1. Follow Steps 1–4 in Soxhlet procedure.
2. Place the thimble in the Goldfish condenser bracket. Push the thimble up so that only about 1/2 in. is below the bracket. Fill the reclaiming beaker with petroleum ether (50 ml) and transfer to beaker. Seal beaker to apparatus using gasket and metal ring. Start the water flow through the condenser. Raise the hotplate up to the beaker, turn on, and start the ether boiling. Extract for 4 hr at a condensation rate of 5–6 drops per second.
3. Follow Steps 6 and 7 in Soxhlet procedure.

Data and Calculations

Using the weights recorded in the tables below, calculate the percent fat (wt/wt) on a wet weight basis as determined by the Soxhlet extraction method. If the fat content of the food you analyzed was given on the label, report this theoretical value.

Name of Snack Food:
Label g fat/serving:
Label serving size (g):
Label g fat/100 g product:

Data from Goldfish extraction:

Trial	Thimble (g)	Wet Sample + Thimble (g)	Wet Sample + Thimble + Glass Wool (g)	Wet Sample (g)	Dry Sample + Thimble + Glass Wool (g)
1					
2					
3					

Data from moisture analysis:

Trial	Pan (g)	Pan + Wet Sample (g)	Pan + Dried Sample (g)	% Moisture
1				
2				
3				
				$\overline{X} =$
				SD $=$

Calculation of % moisture:

$$\frac{\left(\begin{array}{c}\text{wt of wet sample}\\ +\text{ pan}\end{array}\right) - \left(\begin{array}{c}\text{wt of dried sample}\\ +\text{ pan}\end{array}\right)}{(\text{wt of wet sample + pan}) - (\text{wt of pan})} \times 100$$

Trial	% Fat + % Water	Calc. % Fat
1		
2		
3		

Calculation of % fat:

% (fat + moisture)

$$= \frac{\left[\left(\begin{array}{c}\text{Initial wt of sample}\\ +\text{ Thimble}\\ +\text{ Glass wool}\end{array}\right) - \left(\begin{array}{c}\text{Final wt of sample}\\ +\text{ Thimble}\\ +\text{ Glass wool}\end{array}\right)\right] \times 100}{\text{Initial wt of sample}}$$

% fat (wt/wt) = (% fat + moisture) − (% moisture)

(*Note:* Use average % moisture in this calculation)

Questions

1. What would be the advantages of using ethyl ether rather than petroleum ether in a solvent extraction method, such as the Goldfish method?
2. What were the advantages of using the Goldfish extraction method rather than the Soxhlet extraction method?
3. If the fat content measured here differed from that reported on the nutrition label, how might this be explained?

METHOD C: MOJONNIER METHOD

Principle

Fat is extracted with a mixture of ethyl ether and petroleum ether. The extract containing the fat is dried and expressed as percent fat by weight.

The assay uses not only ethyl ether and petroleum ether, but also ammonia and ethanol. Ammonia dissolves the casein and neutralizes the acidity of the product to reduce its viscosity. Ethanol prevents gelation of the milk and ether, and aids in the separation of the ether-water phase. Ethyl ether and petroleum ether serve as lipid solvents, and petroleum ether decreases the solubility of water in the ether phase.

Chemicals

	CAS No.	Hazards
Ammonium hydroxide	1336–21–6	Corrosive, dangerous for the environment
Ethanol	64–17–5	Highly flammable
Petroleum ether	8032–32–4	Harmful, highly flammable, dangerous for environment
(or Ethyl ether)	60–29–7	Harmful, extremely flammable

Hazards, Precautions, and Waste Disposal

Ethanol, ethyl ether, and petroleum ether are fire hazards; avoid open flames, breathing vapors, and contact with skin. Ether is extremely flammable, is hygroscopic, and may form explosive peroxides. Ammonia is a corrosive; avoid contact and breathing vapors. Otherwise, adhere to normal laboratory safety procedures. Wear gloves and safety glasses at all times. Petroleum ether and ether liquid wastes must be disposed of in designated hazardous waste receptacles. The aqueous waste can go down the drain with a water rinse.

Supplies

- Milk, whole and 2%
- Mojonnier extraction flasks, with stoppers
- Mojonnier fat dishes
- Plastic gloves
- Tongs

Equipment

- Analytical balance
- Hot plate
- Mojonnier apparatus (with centrifuge, vacuum oven, and cooling desiccator)

Notes

Reagents must be added to the extraction flask in the following order: water, ammonia, alcohol, ethyl ether, and petroleum ether. The burets on the dispensing cans or tilting pipets are graduated for measuring the proper amount. Make triplicate determinations on both the sample and reagent blanks. The procedure given here is for fresh milk. Other samples may need to be diluted with distilled water in step 2 and require different quantities of reagents in subsequent steps. Consult the instruction manual or AOAC *Official Methods of Analysis* for samples other than fresh milk.

Procedure

(Instructions are given for analysis in triplicate.)

1. Turn on power unit and temperature controls for oven and hot plate on the fat side of the Mojonnier unit.
2. Warm milk samples to room temperature and mix well.
3. When oven is at 135°C, heat cleaned fat dishes in oven under a vacuum of 20 in. mercury for 5 min. Handle dishes from this point on with tongs or gloves. Use three dishes for each type of milk samples, and two dishes for the reagent blank.
4. Cool dishes in cooling desiccator for 7 min.
5. Weigh dishes, record weight of each dish and its identity, and place dishes in desiccator until use.
6. Weigh samples accurately into Mojonnier flasks. If weighing rack is used, fill curved pipettes and place in rack on the balance. Weigh each sample by difference.
7. Add chemicals for the first extraction in the order and amounts given below. After each addition of chemicals, stopper the flask and shake by inverting for 20 sec.

Chemicals	First Extraction		Second Extraction	
	Step	Amount	Step	Amount
Ammonia	1	1.5 ml	—	None
Ethanol	2	10 ml	1	5 ml
Ethyl ether	3	25 ml	2	25 ml
Petroleum ether	4	25 ml	3	25 ml

8. Place the extraction flasks in the holder of the centrifuge. Place both flask holders in the centrifuge. Operate the centrifuge to run at 30 turns in 30 s, to give a speed of 600 rpm (revolutions

per minute). [In lieu of centrifuging, the flasks can be allowed to stand 30 min until a clear separation line forms, or 3 drops of phenolphthalein indicator (0.5% w/v ethanol) can be added during the first extraction to aid in determining the interface.]

9. Carefully pour off the ether solution of each sample into a previously dried, weighed, and cooled fat dish. Most or all of the ether layer should be poured into the dish, but none of the remaining liquid must be poured into the dish.

10. Place dishes with ether extract on hot plate under glass hood of Mojonnier unit, with power unit running. (If this hot plate is not available, use a hot plate placed in a hood, with the hot plate at 100°C.)

11. Repeat the extraction procedure a second time for the samples in the Mojonnier flasks, following the sequence and amount given in the table above. Again, after each addition of chemicals, stopper the flask and shake by inverting for 20 s. Centrifuge the flasks again, as described above. Distilled water now may be added to the flask to bring the dividing line between ether and water layers to the center of neck of flask. If this is done, repeat the centrifugation.

12. Pour ether extract into respective fat dish (i.e., the ether for a specific sample should be poured into the same fat dish used for that sample from the first extraction), taking care to remove all the ether but none of the other liquid in the flask.

13. Complete the evaporation of ether on the hot plate. The ether should boil slowly; not fast enough to cause splattering. If the plate appears to be too hot and boiling is too fast, only part of the dish should be placed on the hot plate. (If the hot plate evaporation of ether is problematic, leave collection containers with lids ajar in an operating hood, to have them evaporated by the next day.)

14. When all the ether has been evaporated from the dishes, place the dishes in the vacuum oven 70–75°C for 10 min. with a vacuum of at least 20 in.

15. Cool the dishes in the desiccator for 7 min.

16. Accurately weigh each dish with fat. Record weight.

Data and Calculations

Calculate the fat content of each sample. Subtract the average weight of the reagent blank from the weight of each fat residue in the calculation.

Trial	Milk Start (g)	Milk End (g)	Milk Tested (g)	Dish (g)	Dish + Fat (g)	Calculated % Fat
Reagent Blank A						—
B						—
					$\overline{X} =$	
Sample 1A						
Sample 1B						
Sample 1C						
						$\overline{X} =$
						SD $=$

$$\% \text{ fat} = 100 \times \{[(\text{wt dish} + \text{fat}) - (\text{wt dish})]$$
$$- (\text{avg wt blank residue})\}/\text{wt sample}$$

Questions

1. List possible causes for high and low results in a Mojonnier fat test.
2. How would you expect the elimination of alcohol from the Mojonnier procedure to affect the results? Why?
3. How would you propose to modify the Mojonnier procedure to test a solid, nondairy product? Explain your answer.

METHOD D: BABCOCK METHOD

Principle

Sulfuric acid is added to a known amount of milk sample in a Babcock bottle. The acid digests the protein, generates heat, and releases the fat. Centrifugation and hot water addition isolate the fat into the graduated neck of the bottle. The Babcock fat test uses a volumetric measurement to express the percent of fat in milk or meat by weight.

Note

The fat column in the Babcock test should be at 57–60°C when read. The specific gravity of liquid fat at that temperature is approximately 0.90 g per ml. The calibration on the graduated column of the test bottle reflects this fact and enables one to make a volumetric measurement which expresses the fat content as percent by weight.

Chemicals

	CAS No.	Hazards
Glymol (red reader)	8042-47-5	Toxic, irritant
Sulfuric acid	7664-93-9	Corrosive

Hazards, Precautions, and Waste Disposal

Concentrated sulfuric acid is an extreme corrosive; avoid contact with skin and clothes and breathing vapors. Wear gloves and safety glasses at all times. Also, wear corrosive resistant gloves. Otherwise, adhere to normal laboratory safety procedures. Sulfuric acid and glymol wastes must be disposed of in a designated hazardous waste receptacle.

For safety and accuracy reasons, dispense the concentrated sulfuric acid from a bottle fitted with a repipettor (i.e., automatic bottle dispenser). Fit the dispenser with a thin, semirigid tube to dispense directly and deep into the Babcock bottle while mixing contents. Set the bottle with dispenser on a tray to collect spills.

Supplies

- 3 Babcock bottles
- Babcock caliber (or shrimp divider)
- Measuring pipette, 10 ml
- Pipette bulb or pump
- Plastic gloves
- Standard milk pipette (17.6 ml)
- Thermometer

Equipment

- Babcock centrifuge
- Water bath

Procedure

(Instructions are given for analysis in triplicate.)

1. Adjust milk sample to ca. 38°C and mix until homogenous. Using a standard milk pipette, pipette 17.6 ml of milk into each of three Babcock bottles. After the pipette has emptied, blow out the last drops of milk from the pipette tip into the bottle. Allow milk samples to adjust to ca. 22°C.
2. Measure ca. 17.5 ml of sulfuric acid (specific gravity 1.82–1.83) and carefully add into the test bottle, with mixing during and between additions, taking care to wash all traces of milk into the bulb of the bottle. Time for complete acid addition should not exceed 20 s. Mix the milk and acid thoroughly. Be careful not to get any of the mixture into the column of the bottle while shaking. Heat generated behind any such lodged mixture may cause a violent expulsion from the bottle.
3. Place bottles in centrifuge heated to 60°C. Be sure bottles are counterbalanced. Position bottles so that bottlenecks will not be broken in horizontal configuration. Be sure the heater of the centrifuge is on.
4. Centrifuge the bottles for 5 min after reaching the proper speed (speed will vary depending upon the diameter of the centrifuge head).
5. Stop the centrifuge and add soft hot water (60°C) until the liquid level is within 0.6 cm of the neck of the bottle. Carefully permit the water to flow down the side of the bottle. Again, centrifuge the bottles for 2 min.
6. Stop the centrifuge and add enough soft hot water (60°C) to bring the liquid column near the top graduation of the scale. Again, centrifuge the bottles for 1 min.
7. Remove the bottles from the centrifuge and place in a heated (55–60°C, preferably 57°C) water bath deep enough to permit the fat column to be below the water level of the water bath. Allow bottles to remain at least 5 min before reading.
8. Remove the samples from the water bath one at a time, and quickly dry the outside of the bottle. Add glymol (red reader) to top of fat layer. Immediately use a divider or caliper to measure the fat column to the nearest 0.05%, holding the bottle in a vertical position at eye level. Measure from the highest point of the upper meniscus to the bottom of the lower meniscus.
9. Reject all tests in which the fat column is milky or shows the presence of curd or charred matter, or in which the reading is indistinct or uncertain. The fat should be clear and sparkling, the upper and lower meniscus clearly defined, and the water below the fat column should be clear.
10. Record the readings of each test and determine the mean % fat and the standard deviation.

Data and Calculations

Trial	Measured % Fat
1	
2	
3	
	$\overline{X} =$
	SD =

Questions

1. What are the possible causes of charred particles in the fat column of the Babcock bottle?
2. What are the possible causes of undigested curd in the Babcock fat test?
3. Why is sulfuric acid preferred over other acids for use in the Babcock fat test?

REFERENCES

AOAC International. 2000. *Official Methods of Analysis*, 17th ed. AOAC International, Gaithersburg, MD.

Min, D.B., and Boff, J.M. 2003. Crude fat analysis. Ch. 8, in *Food Analysis*, 3rd ed. S.S. Nielsen (Ed.), Kluwer Academic, New York.

Wehr, H.M., and Frank, J.F. (Eds.). 2002. *Standard Methods for the Examination of Dairy Products*, 17th ed. American Public Health Administration, Washington, DC.

Protein Nitrogen Determination

INTRODUCTION

Background

The protein content of foods can be determined by numerous methods. The Kjeldahl method and the nitrogen combustion (Dumas) method for protein analysis are based on nitrogen determination. Both methods are official for the purposes of nutrition labeling of foods. While the Kjeldahl method has been used widely for over a hundred years, the recent availability of automated instrumentation for the Dumas method in many cases is replacing use of the Kjeldahl method.

Reading Assignment

Chang, S.K.C. 2003. Protein analysis. Ch. 9, in *Food Analysis*, 3rd ed. S.S. Nielsen (Ed.), Kluwer Academic, New York.

Notes

Both the Kjeldahl and nitrogen combustion methods can be done without automated instrumentation, but are commonly done with automated instruments. The descriptions below are based on the availability of such automated instrumentation. If protein content of samples analyzed by Kjeldahl and/or nitrogen combustion has been estimated in a previous experiment by near infrared analysis, values can be compared between methods.

METHOD A: KJELDAHL NITROGEN METHOD

Objective

Determine the protein content of corn flour using the Kjeldahl method.

Principle of Method

The Kjeldahl procedure measures the nitrogen content of a sample. The protein content then can be calculated assuming a ratio of protein to nitrogen for the specific food being analyzed. The Kjeldahl procedure can be basically divided into three parts: (1) digestion, (2) distillation, (3) titration. In the digestion step, organic nitrogen is converted to an ammonium in the presence of a catalyst at approximately 370°C. In the distillation step the digested sample is made alkaline with NaOH and the nitrogen is distilled off as NH_3. This NH_3 is "trapped" in a boric acid solution. The amount of ammonia nitrogen in this solution is quantified by titration with a standard HCl solution. A reagent blank is carried through the analysis and the volume of HCl titrant required for this blank is subtracted from each determination.

Chemicals

	CAS No.	Hazards
Boric acid (H_3BO_3)	10043-35-3	
Bromocresol green	76-60-8	
Ethanol, 95%	64-17-5	Highly flammable
Hydrochloric acid, conc. (HCl)	7647-01-0	Corrosive
Methyl red	493-52-7	
Sodium hydroxide (NaOH)	1310-73-2	Corrosive
Sulfuric acid, conc. (H_2SO_4)	7664-93-9	Corrosive
Kjeldahl digestion tablets		Irritant
Potassium sulfate (K_2SO_4)	7778-80-5	
Cupric sulfate	7758-98-7	
Titanium dioxide (TiO_2)	13463-67-7	
Tris (hydroxymethyl) aminomethane (THAM)	77-86-1	Irritant

Reagents

(**It is recommended that these solutions be prepared by laboratory assistant before class.)

- Sulfuric Acid (concentrated, N-Free)
- Catalyst/Salt Mixture (Kjeldahl digestion tablets) Contains potassium sulfate, cupric sulfate, and titanium dioxide.
  Note: There are several types of Kjeldahl digestion tablets that contain somewhat different chemicals.
- Sodium Hydroxide Solution, 50% w/v NaOH in deionized distilled (dd) water**
  Dissolve 2000 g sodium hydroxide (NaOH) pellets in ~3.5 L dd water. Cool. Add dd water to make up to 4.0 L.
- Boric Acid Solution**
  In a 4-L flask dissolve 160 g boric acid in ca. 2 L boiled, and still very hot, dd water. Mix and then add an additional 1.5 L of boiled, hot dd water. Cool to room temperature under tap water (caution: glassware may break due to sudden cooling) or leave overnight. When using the rapid procedure, the flask must be shaken occasionally to prevent recrystallization of the boric acid. Add 40 ml of bromocresol green solution (100 mg bromocresol green/100 ml ethanol) and 28 ml of methyl red solution (100 mg methyl red/100 ml ethanol). Dilute to 4 liters with water and mix carefully. Transfer 25 ml of the boric acid solution to a receiver flask and distill a digested blank (a digested catalyst/salt/acid mixture). The contents of the flask should then be a neutral gray. If not, titrate with 0.1 *N* NaOH solution until this color is obtained. Calculate the amount of NaOH

solution necessary to adjust the boric acid solution in the 4-L flask with the formula:

$$\text{ml } 0.1 \, N \text{ NaOH} = \frac{(\text{ml titer}) \times (4000 \text{ ml})}{(25 \text{ml})}$$

Add the calculated amount of 0.1 N NaOH solution to the boric acid solution. Mix well. Verify the adjustment results by distilling a new blank sample. Place adjusted solution into a bottle equipped with a 50-ml repipettor.

- Standardized HCl solution**
  Dilute 3.33 ml conc. HCl to 4 L with dd water. Empty old HCl solution from the titrator reservoir and rinse three times with a small portion of the new HCl solution. Fill the titrator with the new HCl solution to be standardized. Using a volumetric pipet, dispense 10 ml aliquots of the THAM solution prepared as described below into 3 Erlenmeyer flasks (50 ml). Add 3–5 drops indicator (3 parts 0.1% bromocresol green in ethanol to 1 part of 0.2% methyl red in ethanol) to each flask and swirl. Titrate each solution with the HCl solution to a light pink endpoint. Note the acid volume used and calculate the normality as described below.

  Calculation to standardize HCl solution:

  Normality

  $$= \frac{\text{ml THAM} \times \text{THAM Normality}}{\text{average acid volume (AAV)}}$$

  $$= \frac{20 \text{ ml} \times 0.01 \, N}{\text{AAV}}$$

  Write the normality of the standardized HCl solution on the stock container.

- Tris (hydroxymethyl) aminomethane (THAM) Solution −(0.01 N)**
  Place 2 g of THAM in a crucible. Leave in a drying oven (95°C) overnight. Let cool in a desiccator. In a 1-L volumetric flask, dissolve 1.2114 g of oven dried THAM in distilled water. Dilute to volume.

Hazards, Cautions, and Waste Disposal

Concentrated sulfuric acid is an extreme corrosive; avoid breathing vapors and contact with skin and clothes. Concentrated sodium hydroxide is a corrosive. Wear gloves and safety glasses at all times and corrosive resistant gloves. Perform the digestions in an operating hood with an aspirating fume trap attached to the digestion unit. Allow samples to cool in the hood before removing the aspirating fume trap from the digestion unit. Otherwise, adhere to normal laboratory safety

procedures. The waste of combined sulfuric acid and sodium hydroxide has been largely neutralized (check pH to ensure it is pH 3–9), so it can be discarded down the drain with a water rinse. However, for disposing any chemical wastes, follow good laboratory practices outlined by environmental health and safety protocols at your institution.

For safety and accuracy reasons, dispense the concentrated sulfuric acid from a bottle fitted with a repipettor (i.e., automatic dispenser). Fit the dispenser with a thin, semirigid tube to dispense directly into the Kjeldahl tube. Set the bottle with dispenser on a tray to collect spills.

Supplies

(Used by students)

- Corn flour (not dried)
- 5 Digestion tubes
- 5 Erlenmeyer flasks, 250 ml
- Spatula
- Weighing paper

Equipment

- Analytical balance
- Automatic titrator
- Kjeldahl digestion and distillation system

Procedure

(Instructions are given for analysis in triplicate. Follow manufacturer's instructions for specific Kjeldahl digestion and distillation system used. Some instructions given here may be specific for one type of Kjeldahl system.)

I. Digestion

1. Turn on digestion block and heat to appropriate temperature.
2. Accurately weigh approximately 0.1 g corn flour. Record the weight. Place corn flour in digestion tube. Repeat for two more samples.
3. Add 1 catalyst tablet and appropriate volume (e.g., 5 ml) of concentrated sulfuric acid to each tube with corn flour. Prepare duplicate blanks: 1 catalyst tablet + 5 ml sulfuric acid + weigh paper (if weigh paper was added with the corn flour samples).
4. Place rack of digestion tubes on digestion block. Cover digestion block with exhaust system turned on.

5. Let samples digest until digestion is complete. The samples should be clear (but neon green), with no charred material remaining.
6. Take samples off the digestion block and allow to cool with the exhaust system still turned on.
7. Carefully dilute digest with an appropriate volume of dd water. Swirl each tube.

II. Distillation

1. Follow appropriate procedure to start up distillation system.
2. Dispense appropriate volume of boric acid solution into the receiving flask. Place receiving flask on distillation system. Make sure the tube coming from the distillation of the sample is submerged in the boric acid solution.
3. Put sample tube in place, making sure it is seated securely, and proceed with the distillation until completed. In this distillation process, a set volume of NaOH solution will be delivered to the tube and a steam generator will distill the sample for a set period of time.
4. Upon completing distillation of one sample, proceed with a new sample tube and receiving flask.
5. After completing distillation of all samples, follow manufacturer's instructions to shut down the distillation unit.

III. Titration

1. Record the normality of the standardized HCl solution, as determined by the teaching assistant.
2. If using an automated pH meter titration system, follow manufacturer's instructions to calibrate the instrument. Put a magnetic stir bar in the receiver flask and place it on a stir plate. Keep the solution stirring briskly while titrating, but do not let the stir bar hit the electrode. Titrate each sample and blank to an endpoint pH of 4.2. Record volume of HCl titrant used.
3. If using a colorimetric endpoint, put a magnetic stir bar in the receiver flask, place it on a stir plate, and keep the solution stirring briskly while titrating. Titrate each sample and blank with the standardized HCl solution to the first faint gray color. Record volume of HCl titrant used.

Data and Calculations

Calculate the percent nitrogen and the percent protein for each of your duplicate or triplicate corn flour

ssamples, then determine average values. The corn flour sample you analyzed was not a dried sample. Report percent protein results on a wet weight basis (wwb) and on a dry weight basis (dwb). Assume a moisture content of 10% (or use the actual moisture content if previously determined on this corn flour sample). Use 6.25 for the nitrogen to protein conversion factor.

$$\% N = \text{Normality HCl}^* \times \frac{\text{corrected acid vol. (ml)}^{**}}{\text{g of sample}}$$

$$\times \frac{14 \text{ g } N}{\text{mol}} \times 100$$

*Normality is in mol/1000 ml

**Corrected acid vol. = (ml std. acid for sample)
$$- \text{(ml std. for blank)}$$

$$\% \text{ Protein} = \% N \times \text{Protein Factor}$$

	Trial	Sample Wt. (g)	Vol. HCl Titrant (ml)	% Nitrogen	% Protein, wwb	%, Protein, dwb
Blank	1	—		—	—	—
	2	—		—	—	—
			$\overline{X} =$			
Sample	1					
	2					
	3					
					$\overline{X} =$	$\overline{X} =$
					SD =	SD =

Questions

1. If the alkali pump timer on the distillation system was set to deliver 25 ml of 50% NaOH and 7 ml of concentrated H_2SO_4 was used to digest the sample, how many milliliters of the 50% NaOH is actually required to neutralize the amount of sulfuric acid used in the digestion? How would your results have been changed if the alkali pump timer malfunctioned and delivered only 15 ml of the 50% NaOH? (Molarity of conc. $H_2SO_4 = 18$)
2. Could phenolphthalein be used as an indicator in the Kjeldahl titration? Why or why not?
3. Describe the function of the following chemicals used in this determination:
   a. Catalyst pellet
   b. Borate
   c. H_2SO_4
   d. NaOH.
4. Why was it not necessary to standardize the boric acid solution?
5. Explain how the factor used to calculate the percent protein for your product was obtained, and why the protein factors for some other cereal grains (e.g., wheat, oats) differ from that for corn.

6. For each of the disadvantages of the Kjeldahl method, give another protein analysis method that overcomes (at least partially) that disadvantage.

METHOD B: NITROGEN COMBUSTION METHOD

Objective

Determine the protein content of corn flour using the nitrogen combustion method.

Principle of Method

The nitrogen combustion method measures the nitrogen content of a sample. The protein content then is calculated assuming a ratio of protein to nitrogen for the specific food being analyzed. In the assay, the sample is combusted at a high temperature (900–950°C) to release nitrogen gas and other products (i.e., water, other gases). The other products are removed, and the nitrogen is quantitated by gas chromatography using a thermal conductivity detector.

Chemicals

	CAS No.	Hazards
Ethylenediaminetetraacetic acid, disodium salt ($Na_2EDTA \cdot 2H_2O$)	60-00-4	Irritant

(The other chemicals used are specific to each manufacturer for the columns within the instrument.)

Hazards, Cautions, and Waste Disposal

During operation, the front panel of the instrument gets very hot. Check instructions of manufacturer for any other hazards, especially those associated with maintenance of instrument.

Supplies

(For student use)

- Corn flour
- Sample cup

Equipment

- Nitrogen combustion unit

Procedure

Follow manufacturer's instructions for startup, analyzing samples, and shutdown.

Weigh appropriate amount of sample into a tared sample cup on an analytical balance. (Sample weight will be coordinated with sample number in autosampler, if autosampler is used.) Remove sample from balance and prepare for insertion following manufacturer's instructions. If an autosampler is used, the weighed sample must be placed into autosampler in the appropriate slot for the sample number. Repeat this procedure for EDTA standard. Sample and standard should be run in duplicate or triplicate.

Data and Calculations

Record the percent nitrogen content for each of your duplicate or triplicate corn flour samples. Calculate protein content from percent nitrogen data, and determine average percent protein. The corn flour sample you analyzed was not a dried sample. Report percent protein results on a wet weight basis (wwb) and on a dry weight basis (dwb). Assume a moisture content of 10% (or use the actual moisture content if previously determined on this corn flour sample). Use 6.25 for the nitrogen to protein conversion factor.

Sample	% Nitrogen	% Protein, wwb	% Protein, dwb
1			
2			
3			
		$\overline{X} =$	$\overline{X} =$
		SD =	SD =

Questions

1. What are the advantages of the nitrogen combustion method compared to the Kjeldahl method?
2. Explain why ethylenediaminetetraacetic acid (EDTA) can be used as a standard to check the calibration of the nitrogen analyzer.
3. If you analyzed the corn flour sample by both the Kjeldahl and nitrogen combustion methods, compare the results. What might explain any differences?

REFERENCES

Chang, S.K.C. 2003. Protein analysis. Ch. 9, in *Food Analysis*, 3rd ed. S.S. Nielsen (Ed.), Kluwer Academic, New York.

AOAC International. 2000. *Official Methods of Analysis*, 17th ed. Method 960.52 (Micro-Kjeldahl Method) and Method 992.23 (Generic Combustion Method). AOAC International, Gaithersburg, MD.

Phenol-Sulfuric Acid Method for Total Carbohydrates

INTRODUCTION

Background

The phenol-sulfuric acid method is a simple and rapid colorimetric method to determine total carbohydrates in a sample. The method detects virtually all classes of carbohydrates, including mono-, di-, oligo-, and polysaccharides. Although the method detects almost all carbohydrates, the absorptivity of the different carbohydrates varies. Thus, unless a sample is known to contain only one carbohydrate, the results must be expressed arbitrarily in terms of one carbohydrate.

In this method, the concentrated sulfuric acid breaks down any polysaccharides, oligosaccharides, and disaccharides to monosaccharides. Pentoses (5-carbon compounds) then are dehydrated to furfural, and hexoses (6-carbon compounds) to hydroxymethyl furfural. These compounds then react with phenol to produce a yellow-gold color. For products that are very high in xylose (a pentose), such as wheat bran or corn bran, xylose should be used to construct the standard curve for the assay, and measure the absorption at 480 nm. For products that are high in hexose sugars, glucose is commonly used to create the standard curve, and the absorption is measured at 490 nm. The color for this reaction is stable for several hours, and the accuracy of the method is within ±2% under proper conditions.

Carbohydrates are the major source of calories in soft drinks, beer and fruit juices, supplying 4 Cal/gram carbohydrate. In this experiment, you will create a standard curve with a glucose standard solution, use it to determine the carbohydrate concentration of soft drinks and beer, then calculate the caloric content of those beverages.

Reading Assignment

BeMiller, J.N. 2003. Carbohydrate analysis. Ch. 10, in *Food Analysis*, 3rd ed. S.S. Nielsen (Ed.), Kluwer Academic, New York.

Objective

Determine the total carbohydrate content of soft drinks and beers.

Principle of Method

Carbohydrates (simple sugars, oligosaccharides, polysaccharides, and their derivatives) react in the presence of strong acid and heat to generate furan derivatives that condense with phenol to form stable yellow-gold compounds that can be measured spectrophotometrically.

Chemicals

	CAS No.	Hazards
D-Glucose ($C_6H_{12}O_6$)	50–99–7	
Phenol (C_6H_6O)	108–95–2	Toxic
Sulfuric acid (H_2SO_4)	7664–93–9	Corrosive

Reagents

(**It is recommend that these solutions be prepared by the laboratory assistant before class.)

- Glucose std solution, 100 mg/L**
- Phenol, 80% wt/wt in H_2O, 1 ml**
    Prepare by adding 20 g deionized distilled (dd) water to 80 g of redistilled reagent grade phenol (crystals)
- Sulfuric acid, concentrated

Hazards, Cautions, and Waste Disposal

Use concentrated H_2SO_4 and the 80% phenol solution with caution. Wear gloves and safety glasses at all times, and use good lab technique. The concentrated H_2SO_4 is very corrosive (e.g., to clothes, shoes, skin). The phenol is toxic and must be discarded as a hazardous waste. Other waste not containing phenol likely may be put down the drain using a water rinse, but follow good laboratory practices outlined by environmental health and safety protocols at your institution.

Supplies

(Used by students)

- Beer (lite and regular, of same brand)
- Bottle to collect waste
- Cuvettes (tubes) for spectrophotometer
- Erlenmeyer flask, 100 ml, for dd water
- 2 Erlenmeyer flasks, 500 ml, for beverages
- Gloves
- Mechanical, adjustable volume pipettors, 1000 and 100 μl (or 200 μl), with plastic tips
- Pasteur pipettes and bulb
- Parafilm®
- Pipette bulb or pump
- Repipettor (for fast-delivery of 5 ml conc. H_2SO_4)
- Soft drinks (clear-colored, diet and regular, of same brand)
- 20 Test tubes, 16–20 mm internal diameter
- Test tube rack
- 4 Volumetric flasks, 100 ml *or* 2 volumetric flasks, 1000 ml

- Volumetric pipette, 5 ml
- 2 Volumetric pipettes, 10 ml

Equipment

- Spectrophotometer
- Vortex mixer
- Water bath, maintained at 25°C

PROCEDURE

(Instructions are given for analysis in duplicate.)

1. Standard curve tubes: Using the glucose standard solution (100 mg glucose/L) and dd water as indicated in the table below, pipet aliquots of the glucose standard into clean test tubes (duplicates for each concentration) such that the tubes contain 0–100 µl of glucose (use 1000 µl mechanical pipettor to pipet samples), in a total volume of 2 ml. These tubes will be used to create a standard curve, with values of 0–100 µg glucose/ 2 ml. The 0 µg glucose/2 ml sample will be used to prepare the reagent blank.

	µg Glucose/2 ml					
	0	20	40	60	80	100
ml glucose stock solution	0	0.2	0.4	0.6	0.8	1.0
ml dd water	2.0	1.8	1.6	1.4	1.2	1.0
ug/ml =	0	10	20	30	40	50

2. Record caloric content from label: You will analyze for total carbohydrate content: (1) a regular and diet soft drink of the same brand, *or* (2) a regular and lite beer of the same brand. Before you proceed with the sample preparation and analysis, record the caloric content on the nutrition label of the samples you will analyze.

3. Decarbonate the beverages: With the beverages at room temperature, pour approximately 100 ml into a 500-ml Erlenmeyer flask. Shake gently at first (try not to foam the sample if it is beer) and continue gentle shaking until no observable carbon dioxide bubbles appear. If there is any noticeable suspended material in the beverage, filter the sample before analysis.

4. Sample tubes: So the sample tested will contain 20–100 µg glucose/2 ml, the dilution procedure and volumes to be assayed are given below. After dilution as indicated, pipette 1.0 ml of sample into a test tube and add 1.0 ml of dd water. Analyze each diluted sample in duplicate.

	Dilution	Volume Assayed (ml)
Soft drink		
Regular	1 : 2000	1
Diet	0	1
Beer		
Regular	1 : 2000	1
Lite	1 : 1000	1

Recommended dilution scheme for 1 : 2000 dilution:
a. Pipette 5 ml of beverage into a 100-ml volumetric flask, and dilute to volume with dd water. Seal flask with Parafilm® and mix well (this is a 1 : 20 dilution). Then, pipette 1.0 ml of this 1 : 20 diluted beverage into another 100-ml volumetric flask. Dilute to volume with dd water. Seal flask with Parafilm® and mix well.
   or
b. Pipette 1.0 ml of beverage into a 1000-ml volumetric flask, and dilute to volume with dd water. Seal flask with Parafilm® and mix well. Then, in a test tube, combine 1 ml of the 1 : 1000 diluted beverage and 1 ml dd water. Mix well.

Recommended dilution scheme for 1 : 1000 dilution:
a. Pipette 10 ml of beverage into a 100-ml volumetric flask, and dilute to volume with dd water. Seal flask with Parafilm® and mix well (this is a 1 : 10 dilution). Then, pipette 1.0 ml of this 1 : 10 diluted beverage into another 100-ml volumetric flask. Dilute to volume with dd water. Seal flask with Parafilm® and mix well.
   or
b. Pipette 1.0 ml of beverage into a 1000-ml volumetric flask, and dilute to volume with dd water. Seal flask with Parafilm and mix well.

5. Phenol addition: To each tube from Parts 1 and 4 containing a total volume of 2 ml, add 0.05 ml 80% phenol (use 100 or 200 µl mechanical pipettor). Mix on a Vortex test tube mixer.

6. H_2SO_4 addition: To each tube from Part 5, add 5.0 ml H_2SO_4. The sulfuric acid reagent should be added rapidly to the test tube. Direct the stream of acid against the liquid surface rather than against the side of the test tube in order to obtain good mixing. (These reactions are driven by the heat produced upon the addition of H_2SO_4 to an aqueous sample. Thus, the rate of addition of sulfuric acid must be standardized.) Mix on a Vortex test tube mixer. Let tubes stand for 10 min and then place in a 25°C bath for 10 min (i.e., to cool them to room temperature). Vortex the test tubes again before reading the absorbance.

7. Reading absorbance: Wear gloves to pour samples from test tubes into cuvettes. Do not rinse cuvettes with water between samples. Zero the spectrophotometer with the standard curve sample that contains 0 μg glucose/2 ml (i.e., blank). Retain this blank sample in one cuvette for later use. Read absorbances of all other samples at 490 nm. Read your standard curve tubes from low to high concentration (i.e., 20 μg/2 ml up to 100 μg/2 ml), and then read your beverage samples. To be sure that the outside of the cuvettes are free of moisture and smudges, wipe the outside of the cuvette with a clean paper wipe prior to inserting it into the spectrophotometer for a reading.

8. Absorbance spectra: Use one of the duplicate tubes from a standard curve sample with an absorbance reading of 0.5–0.8. Determine the absorbance spectra from 450–550 nm by continually zeroing the spectrophotometer with the blank, then reading the tube at 10 nm intervals.

DATA AND CALCULATIONS

1. Summarize your procedures and results for all standards and samples in the table immediately below. One standard curve sample and one soft drink sample are shown as examples in the second table below. Note that for the example soft drink sample, the μg glucose in the tube and the g glucose/L were calculated using an example equation of the line for the standard curve, and taking into account the dilution and volume assayed.

Tube #	Sample Identity	Dilution Scheme	ml Diluted Std or Unknown	A_{490}	Glucose Equivalent μg in Tube	Glucose Equivalent g/L in Original Sample
1	Blank					
2	Std. 20 μg					
3	Std. 20 μg					
4	Std. 40 μg					
5	Std. 40 μg					
6	Std. 60 μg					
7	Std. 60 μg					
8	Std. 80 μg					
9	Std. 80 μg					
10	Std. 100 μg					
11	Std. 100 μg					
12	Soft drink, reg.					

Tube #	Sample Identity	Dilution Scheme	ml Diluted Std or Unknown	A_{490}	Glucose Equivalent μg in Tube	Glucose Equivalent g/L in Original Sample
13	Soft drink, reg.					
14	Soft drink, diet					
15	Soft drink, diet					
16	Beer, reg.					
17	Beer, reg.					
18	Beer, lite					
19	Beer, lite					

Sample table:

Tube #	Sample Identity	Dilution Scheme	ml Diluted Std or Unknown	A_{490}	Glucose Equivalent μg in Tube	Glucose Equivalent g/L in Original Sample
2	Std, 20 μg	—	0.2 ml	0.243	20	0.10
18	Soft drink, regular	1:2000	1 ml	0.648	79	158

Sample calculation for soft drink, regular:

Equation of the line: $y = 0.011x + 0.1027$

$y = 0.648$

$x = 78.98\ \mu g/2ml$

$(78.98\ \mu g\ glucose/2\ ml) \times (2\ ml/1\ ml)$

$\times (2000\ ml/1\ ml) = 157950\ \mu g/ml$

$= 157.95\ mg/ml$

$= 157.95\ g/L$

2. Construct a standard curve for your total carbohydrate determinations, expressed in terms of glucose (A_{490} versus μg glucose/2 ml). Determine the equation of the line for the standard curve.

3. Calculate the concentration of glucose in your soft drink samples and beer samples, in terms of (a) grams/liter, and (b) g/12 fl. oz. (*Note:* 29.56 ml/fl. oz.)

4. Calculate the caloric content (based only on carbohydrate content) of your soft drink samples and beer samples in term of Cal/12 fl. oz.

Sample	g Glucose/ 12 fl. oz.	Measured Cal/12 fl. oz.	Nutrition Label Cal/12 fl. oz.
Soft drink			
Regular			
Diet			
Beer			
Regular			
Lite			

5. Plot the absorbance spectra obtained by measuring the absorbance between 450 and 550 nm.

nm	450	460	470	480	490	500	510	520	530	540	550
Abs.											

QUESTIONS

1. What are the advantages, disadvantages, and sources of error for this method to determine total carbohydrates?
2. Your lab technician performed the phenol-H_2SO_4 analysis on food samples for total carbohydrates but the results showed low precision, and the values seemed a little high. The technician had used new test tubes (they had never been used, and were taken right from the cardboard box). What most likely caused these results? Why? Describe what happened.
3. If you started with a glucose standard solution of 10 g glucose/liter, what dilution of this solution would be necessary such that you could pipette 0.20, 0.40, 0.60, 0.80, 1.0 ml of the diluted glucose standard solution into test tubes and add water to 2 ml for the standard curve tubes (20–100 μg/2 ml)? Show all calculations.
4. If you had not been told to do a 2000-fold dilution of a soft drink sample, and if you know the approximate carbohydrate content of regular soft drinks (U.S. Department of Agriculture Nutrient Database for Standard Reference indicates ca. 3 g carbohydrate/fl. oz.), how could you have calculated the 2000-fold dilution was appropriate if you wanted to use 1 ml of diluted soft drink in the assay. Show all calculations.
5. How does your calculated value compare to the caloric content on the food label? Do the rounding rules for Calories explain any differences? (See Tables 3–5 of Nielsen, *Food Analysis*.) Does the alcohol content (assume 4–5% alcohol at 7 Cal/g) of beer explain any differences?
6. Was it best to have read the absorbance for the standard curve and other samples at 490 nm? Explain why a wavelength in this region is appropriate for this reaction.

ACKNOWLEDGMENT

This laboratory was developed with input from Dr Joseph Montecalvo, Jr., Department of Food Science & Nutrition, California Polytechnic State University, San Luis Obispo, CA.

REFERENCES

BeMiller, J.N. 2003. Carbohydrate analysis. Ch. 10, in *Food Analysis*, 3rd ed. S.S. Nielsen (Ed.), Kluwer Academic, New York.

Dubois, M., Gilles, K.A., Hamilton, J.K., Rebers, P.A., and Smith, F. 1956. Colorimetric method for determination of sugars and related substances. *Analytical Chemistry* 28: 350–356.

7 chapter

Vitamin C Determination by Indophenol Method

INTRODUCTION

Background

Vitamin C is an essential nutrient in the diet, but is easily reduced or destroyed by exposure to heat and oxygen during processing, packaging, and storage of food. The U.S. Food and Drug Administration requires the Vitamin C content to be listed on the nutrition label of foods. The instability of Vitamin C makes it more difficult to ensure an accurate listing of Vitamin C content on the nutrition label.

The official method of analysis for Vitamin C determination of juices is the 2,6-dichloroindophenol titrimetric method (AOAC Method 967.21). While this method is not official for other types of food products, it is sometime used as a rapid, quality control test for a variety of food products, rather than the more time-consuming microfluorometric method (AOAC Method 984.26). The procedure outlined below is from AOAC Method 967.21.

Reading Assignment

AOAC International. 2000. *Official Methods of Analysis.* 17th ed. Method 967.21. AOAC International, New York.

Eitenmiller, R.R., and Landen, W.O. 2003. Vitamin analysis. Ch. 11, in *Food Analysis*, 3rd ed. S.S. Nielsen (Ed.), Kluwer Academic, Gaithersburg, MD.

Objective

Determine the Vitamin C content of various orange juice products using the indicator dye 2,6-dichloro-indophenol in a titration method.

Principle of Method

Ascorbic acid reduces the indicator dye to a colorless solution. At the endpoint of titrating an ascorbic acid-containing sample with dye, excess unreduced dye is a rose-pink color in the acid solution. The titer of the dye can be determined using a standard ascorbic acid solution. Food samples in solution then can be titrated with the dye, and the volume for the titration used to calculate the ascorbic acid content.

Chemicals

	CAS No.	Hazards
Acetic acid (CH_3COOH)	64-19-7	Corrosive
Ascorbic acid	50-81-7	
2,6-Dichloroindophenol (DCIP)(sodium salt)	620-45-1	
Metaphosphoric acid (HPO_3)	37267-86-0	Corrosive
Sodium bicarbonate ($NaHCO_3$)	144-55-8	

Reagents

(**It is recommended that samples be prepared by the laboratory assistant before class.)

- Ascorbic acid standard solution (prepare only at time of use)
  Accurately weigh (on an analytical balance) approximately 50 mg ascorbic acid [preferably U.S. Pharmacopia (USP) Ascorbic Acid Reference Standard]. Record this weight. Transfer to a 50-ml volumetric flask. Dilute to volume *immediately before use* with the metaphosphoric acid–acetic acid solution (see below for preparation of this solution).

- Indophenol solution—dye
  To 50 ml deionized distilled (dd) water in a 150-ml beaker, add and stir to dissolve 42 mg sodium bicarbonate, then add and stir to dissolve 50 mg 2,6-dichloroindophenol sodium salt. Dilute mixture to 200 ml with dd water. Filter through fluted filter paper into an amber bottle. Close the bottle with a stopper or lid and store refrigerated until used.

- Metaphosphoric acid–acetic acid solution
  To a 250-ml beaker, add 100 ml dd water then 20 ml acetic acid. Add and stir to dissolve 7.5 g metaphosphoric acid. Dilute mixture to 250 ml with distilled water. Filter through fluted filter paper into a bottle. Close the bottle with a stopper or lid and store refrigerated until used.

- Orange juice samples**
  Use products processed and packaged in various ways (e.g., canned, reconstituted frozen concentrate, fresh squeezed, not-from-concentrate). Filter juices through cheesecloth to avoid problems with pulp when pipetting. Record from the nutrition label for each product the percent of the Daily Value for Vitamin C.

Hazards, Precautions, and Waste Disposal

Preparation of reagents involves corrosives. Use appropriate eye and skin protection. Otherwise, adhere to normal laboratory safety procedures. Waste likely may be put down the drain using a water rinse, but follow good laboratory practices outlined by environmental health and safety protocols at your institution.

Supplies

(Used by students)

- Beaker, 150 ml
- Beaker, 250 ml
- 2 Bottles, glass, 200–250 ml, one amber and one clear, both with lids or stoppers

- Buret, 50 or 25 ml
- 9 Erlenmeyer flasks, 50 ml (or 125 ml)
- Fluted filter paper, 2 pieces
- Funnel, approx. 6–9 cm diameter (to hold filter paper)
- Funnel, approx. 2–3 cm diameter (to fill buret)
- 2 Glass stirring rods
- Graduated cylinder, 25 ml
- Graduated cylinder, 100 ml
- Pipette bulb or pump
- Ring stand
- 3 Spatulas
- Volumetric flask, 50 ml
- Volumetric flask, 200 ml
- Volumetric flask, 250 ml
- 2 Volumetric pipettes, 2 ml
- Volumetric pipette, 5 ml
- Volumetric pipette, 7 ml
- Volumetric pipette, 10 or 20 ml
- Weighing boats or paper

Equipment

- Analytical balance

Notes

The instructor may want to assign one or two types of orange juice samples to each student (or lab group) for analysis, rather than having all students analyze all types of orange juice samples. Quantities of supplies and reagents specified are adequate for each student (or lab group) to standardize the dye and analyze one type of orange juice sample in triplicate.

PROCEDURE

(Instructions are given for analysis in triplicate.)

Standardization of Dye

1. Pipette 5 ml metaphosphoric acid–acetic acid solution into each of three 50-ml Erlenmeyer flasks.
2. Add 2.0 ml ascorbic acid standard solution to each flask.
3. Using a funnel, fill the buret with the indophenol solution (dye) and record the initial buret reading.
4. Place the Erlenmeyer flask under the tip of the buret. Slowly add indophenol solution to standard ascorbic acid solution until a light but distinct rose-pink color persists for >5 s (takes about 15–17 ml). Swirl the flask as you add the indophenol solution.

5. Note final buret reading and calculate the volume of dye used.
6. Repeat steps 3–5 for the other two standard samples. Record the initial and final buret readings and calculate the volume of dye used for each sample.
7. Prepare blanks: Pipette 7.0 ml metaphosphoric acid–acetic acid solution into each of three 50-ml Erlenmeyer flasks. Add to each flask a volume of distilled water approximately equal to the volume of dye used above (i.e., average volume of dye used to titrate three standard samples).
8. Titrate the blanks in the same way as steps 3–5 above. Record initial and final buret readings for each titration of the blank, and calculate the volume of dye used.

Analysis of Juice Samples

1. Pipet into each of three 50-ml Erlenmeyer flasks 5 ml metaphosphoric acid–acetic acid solution and 2 ml orange juice.
2. Titrate each sample with the indophenol dye solution (as you did in Steps 3–5 above) until a light but distinct rose-pink color persists for >5 s.
3. Record the initial and final readings and calculate the difference to determine the amount of dye used for each titration.

DATA AND CALCULATIONS

Data

	Trial	Buret Start (ml)	Buret End (ml)	Vol. Titrant (ml)
Ascorbic acid standards	1			
	2			
	3			
				$\overline{X} =$
Blank	1			
	2			
	3			
				$\overline{X} =$
Sample	1			
	2			
	3			

Calculations

1. Using the data obtained in standardization of the dye, calculate the titer using the following

formula:

Titer = F

$$= \frac{\text{mg ascorbic acid in volume of standard solution titrated**}}{\text{(average ml dye used to titrate standards)} - \text{(average ml dye used to titrate blank)}}$$

**mg ascorbic acid in volume of standard solution titrated

$$= (\text{mg of ascorbic acid}/50 \text{ ml}) \times 2 \text{ ml}$$

2. Calculate the ascorbic acid content of the juice sample in mg/ml, using the equation that follows and the volume of titrant for each of your replicates. Calculate the mean and standard deviation of the ascorbic acid content for your juice (in mg/ml). Obtain from other lab members the mean ascorbic acid content (in mg/ml) for other types of juice. Use these mean values for each type of juice to express the Vitamin C content of the juice samples as milligram ascorbic acid/100 ml, and as milligram ascorbic acid/8 fl. oz. (29.56 ml/fl. oz.).

$$\text{mg ascorbic acid}/\text{ml} = (X - B) \times (F/E) \times (V/Y)$$

where:
    X = average ml for sample titration
    B = average ml for sample blank titration
    F = titer of dye (= mg ascorbic acid equivalent to 1.0 ml indophenol standard solution)
    E = ml assayed (= 2 ml)
    V = volume of initial assay solution (= 7 ml)
    Y = volume of sample aliquot titrated (= 7 ml)

Ascorbic acid (AA) content for replicates of orange juice sample:

Replicate	mg AA/ml
1	
2	
3	
$\overline{X}$ =	
SD =	

Summary of ascorbic acid (AA) content of orange juice samples:

Sample Identity	mg AA/ml	mg AA/100 ml	mg AA/8 fl. oz.
1			
2			
3			
4			

Example calculation:

Weight of ascorbic acid used = 50.2 mg

Average volume of titrant used:
    Ascorbic acid standards = 15.5 ml
    Blanks = 0.10 ml

Volume of titrant used for orange juice sample = 7.1 ml

$$\text{Titer} = F = \frac{[(50.2 \text{ mg}/50 \text{ ml}) \times 2 \text{ ml}]}{(15.5 \text{ ml} - 0.10 \text{ ml})}$$

$$= 0.130 \text{ mg}/\text{ml}$$

$$\text{mg ascorbic acid}/\text{ml} = (7.1 \text{ ml} - 0.10 \text{ ml})$$
$$\times (0.130 \text{ mg}/2 \text{ ml})$$
$$\times (7 \text{ ml}/7 \text{ ml})$$
$$= 0.455 \text{ mg}/\text{ml}$$

$$0.455 \text{ mg}/\text{ml} = 45.4 \text{ mg}/100 \text{ ml}$$

0.455 mg ascorbic acid/ml juice
    $\times$ 29.56 ml/fl. oz. $\times$ 8 fl. oz.
    = 107.6 ml ascorbic acid/8 fl. oz.

QUESTIONS

1. By comparing results obtained for various orange juice products, did heat and/or oxygen exposure during processing and storage of the samples analyzed seem to affect the Vitamin C content?
2. How do results you have available for the juice samples analyzed compare to: (1) values listed on the nutrition label for the same juice product, and (2) values in the U.S. Department of Agriculture Nutrient Database for Standard Reference (Web address: http://www.nal.usda.gov/fnic/foodcomp). For the nutrition label values, convert percent of Daily Value to

mg/8 fl. oz., given that the Daily Value for Vitamin C is 60 mg. Why might the Vitamin C content determined for a specific orange juice product not match the value as calculated from percent of Daily Value on the nutrition label?

Ascorbic acid content of orange juices (mg AA/8 fl. oz.):

Sample Identity	Lab Values	USDA Database	Nutrition Label
1			
2			
3			
4			

3. Why was it necessary to standardize the indophenol solution?
4. Why was it necessary to titrate blank samples?
5. Why might the Vitamin C content as determined by this method be underestimated in the case of the heat processed juice samples?

REFERENCES

AOAC International. 2000. Method 967.21, *Official Methods of Analysis*. 17th ed. AOAC International, Gaithersburg, MD.

Eitenmiller, R.R., and Landen, W.O. 2003. Vitamin analysis. Ch. 11, in *Food Analysis*, 3rd ed. S.S. Nielsen (Ed.), Kluwer Academic, New York.

Complexometric Determination of Calcium

INTRODUCTION

Background

Ethylenediaminetetraacetate (EDTA) complexes with numerous mineral ions, including calcium and magnesium. This reaction can be used to determine the amount of these minerals in a sample by a complexometric titration. Endpoints in the titration are detected using indicators that change color when they complex with mineral ions. Calmagite and eriochrome black T (EBT) are such indicators that change from blue to pink when they complex with calcium and magnesium. In the titration of a mineral-containing solution with EDTA, with either indicator the solution turns from pink to blue at the endpoint. The pH affects a complexometric EDTA titration in several ways, and must be carefully controlled. A major application of EDTA titration is testing the hardness of water, for which the method described is an official method (Standard Methods for the Examination of Water and Wastewater, Method 2340C; AOAC Method 920.196).

Hardness of water also can be tested by a more rapid test strip method. Such test strips are available from various companies. The strips contain EDTA and an indicator chemical to cause a color change when calcium and magnesium in water react with the EDTA.

Reading Assignment

Carpenter, C.E., and Hendricks, D.G. 2003. Mineral analysis. Ch. 12, in *Food Analysis*, 3rd ed. S.S. Nielsen (Ed.), Kluwer Academic, New York.

Objective

Determine the hardness of water by EDTA titration and with Quantab® test strips.

METHOD A: EDTA TITRIMETRIC METHOD FOR HARDNESS OF WATER

Principle of Method

Ethylenediaminetetraacetic acid (EDTA) forms a stable 1:1 complex with calcium or magnesium at pH 10. The metal ion indicators, Calmagite and eriochrome black T (EBT), are pink when complexed to metal ions but blue when no metal ions are complexed to them. The indicators bind to metal ions less strongly than does EDTA. When the indicator is added to a solution containing metal ions, the solution becomes pink. When EDTA is added as titrant to the mineral-containing sample, metal ions preferentially complex with the EDTA, leaving the indicator without a metal ion complexed.

When enough EDTA has been titrated to complex with all the metal ions present, the indicator appears blue. This blue color is the endpoint of the titration. The volume and concentration of the EDTA used in the titration are used to calculate the concentration of calcium in the sample, expressed as mg calcium carbonate/liter. Stoichiometry of the reaction is 1 mol of calcium complexing with 1 mol of EDTA.

Chemicals

	CAS No.	Hazards
Ammonium chloride (NH_4Cl)	12125-02-9	Harmful
Ammonium hydroxide (NH_4OH)	1336-21-6	Corrosive, dangerous for the environment
Calcium carbonate ($CaCO_3$)	471-34-1	
Calmagite[3-hydroxy-4-(6-hydroxy-*m*-tolylazo) naphthalene-1-sulfonic acid]	3147-14-6	
Ethylenediaminetetraacetic acid, disodium salt ($Na_2EDTA \cdot 2H_2O$)	60-00-4	Irritant
Hydrochloric acid, concentrated (HCl)	7647-01-0	Corrosive
Magnesium chloride, hexahydrate ($MgCl_2 \cdot 6H_2O$)	7791-18-6	
Magnesium sulfate, heptahydrate ($MgSO_4 \cdot 7H_2O$)	10034-99-8	

Reagents

(**It is recommended that these solutions be prepared by the laboratory assistant before class.)

- Buffer solution**
  Dissolve 16.9 g NH_4Cl in 143 ml concentrated NH_4OH. Dissolve 1.179 g $Na_2EDTA \cdot 2H_2O$ (analytical reagent grade) and 780 mg $MgSO_4 \cdot 7H_2O$ or 644 g $MgCl_2 \cdot 6H_2O$ in 50 ml deionized distilled (dd) water. Combine these two solutions with mixing and dilute to 250 ml with dd water. Store in tightly stoppered Pyrex or plastic bottle to prevent loss of ammonia (NH_3) or pickup of carbon dioxide (CO_2). Dispense this buffer solution with a repipette system. Discard buffer when 1–2 ml added to a sample fails to give pH 10.0 ± 0.1 at the endpoint of the titration.

- Calcium standard solution, 1.000 mg $CaCO_3$/ml** (modified from official method; omit use of methyl red indicator)
  Use primary standard or special reagent low in heavy metals, alkalis, and magnesium. Dry $CaCO_3$ at 100°C for 24 hr. Accurately weigh ca. 1.0 g $CaCO_3$. Transfer to a 500-ml Erlenmeyer flask. Place a funnel in the neck of the flask and add, a little at a time, HCl (1:1, conc. HCl : H_2O), until all $CaCO_3$ has dissolved (be sure all $CaCO_3$ in neck of flask has been washed down with HCl). Add 200 ml dd water and boil a few minutes to expel CO_2. Cool. Adjust to pH 3.8 with 3 M NH_4OH or HCl (1:1, conc. HCl : H_2O), as required. Transfer quantitatively to a 1-L volumetric flask, and dilute to volume with dd water. (1 ml = 1.00 mg $CaCO_3$)
- EDTA standard solution, 0.01 M
  Weigh 3.723 g $Na_2EDTA \cdot 2H_2O$. Dilute to 1 L with dd water. Store in polyethylene (preferable) or borosilicate glass bottles. Standardize this solution using the calcium standard solution as described in the Procedure.
- Hydrochloric acid, 1:1 with water**
  To 10 ml of dd water, add 10 ml concentrated HCl. Mix carefully.
- Calmagite
  Dissolve 0.10 g Calmagite in 100 ml dd water. Use 1 ml per 30 ml solution to be titrated. Put in bottle with eye dropper.

Notes

In this experiment, Calmagite will be used as the indicator dye rather than EBT. Unlike EBT, Calmagite is stable in aqueous solution. Calmagite produces the same color change as EBT, but with a sharper endpoint.

To give a satisfactory endpoint, magnesium ions must be present. To ensure this, a small amount of neutral magnesium salt is added to the buffer.

The specified pH of 10.0 + 0.1 is a compromise situation. With increasing pH, the sharpness of the endpoint increases. However, at high pH, the indicator dye changes color and there is risk of precipitating calcium carbonate ($CaCO_3$) or magnesium hydroxide. The tendency toward $CaCO_3$ precipitation is the reason for the titration duration time limit of 5 min.

Fading or indistinct endpoints can be caused by interference from some metal ions. Certain inhibitors can be added before titration to reduce this interference, but the inhibitors specified are toxic (i.e., sodium cyanide) or malodorous. Magnesium salt of 1,2-cyclohexanediaminetetraacetic acid (MgCDTA), which selectively complexes heavy metals, may be substituted for these inhibitors. However, for samples with high concentrations of heavy metals, a non-EDTA method is recommended. In this experiment, inhibitors or MgCDTA will not be used.

Hazards, Precautions, and Waste Disposal

Adhere to normal laboratory safety procedures. Wear gloves and safety glasses at all times. The buffer solution, which contains ammonium hydroxide, should be disposed of as a hazardous waste. Other wastes likely may be put down the drain using a water rinse, but follow good laboratory practices outlined by environmental health and safety protocols at your institution.

Supplies

(Used by students)

- Buret, 25 or 50 ml
- 9 Erlenmeyer flasks, 125 ml
- Funnel (to fill buret)
- 3 Graduated cylinders, 25 ml
  (Graduated cylinder of larger volumes may be necessary; for example, 100 ml or larger; size to be determined by trial in Procedure II.1.)
- Mechanical pipettor, 1000 μl, with plastic tips
- Pasteur pipette and bulb
- Volumetric flask, 1000 ml
- Volumetric pipette, 10 ml
- Weighing paper/boat

Equipment

- Analytical balance
- Drying oven, 100°C
- Hot plate
- pH meter

Procedure

(Modified from Method 2340 Hardness, *Standard Methods for the Examination of Water and Wastewater*, 20th ed.) (Instructions are given for analysis in triplicate.)

I. Standardization of EDTA Solution

1. Pipette 10 ml of calcium standard solution into each of three 125-ml Erlenmeyer flasks.
2. Adjust to pH 10.0 ± 0.05 with buffer solution. (If possible, do this pH adjustment with the buffer in an operating hood, due to its odor.) As necessary, use the HCl solution (1:1) in pH adjustment.
3. Titrate each flask with EDTA solution slowly, with continuous stirring, until last reddish tinge disappears, adding last few drops at 3–5 s intervals. Color at endpoint is blue in daylight and under daylight fluorescent lamp. Color may first appear lavender or purple, but will then turn to blue. Complete titration within 5 min from time of buffer addition.

4. Record the volume of EDTA solution used for each titration.

II. Titration of Water Sample

1. Dilute 25-ml tap water sample (or such volume as to require <15 ml titrant) to ca. 50 ml with dd water in 125-ml Erlenmeyer flask. For tap distilled water, test 50 ml, without dilution. Prepare samples in triplicate. [Official method recommends the following: For water of low hardness (<5 mg/L), use 100–1000 ml specimen, proportionately larger amounts of reagents, microburet, and blank of distilled water equal to specimen volume.]
2. Adjust pH to 10 ± 0.05 as described in Step I.2.
3. Titrate each sample with EDTA standard solution slowly, as described in Step I.3 above for standardization of EDTA solution.
4. Record the volume of EDTA solution used for each titration.

Data and Calculations

Calculate molarity of calcium standard solution:

$$g \, CaCO_3 =$$

Molarity of calcium solution

$$= \frac{g \, CaCO_3}{(100.09 \, g/mol)(liter \, solution)}$$

$$= mol \, calcium/L$$

Standardization of EDTA solution:

Trial	Buret Start (ml)	Buret End (ml)	Volume Titrant (ml)	Molarity
1				
2				
3				
			$\overline{X} =$	
			$SD =$	

Calculate molarity of EDTA solution:

$$mol \, calcium = mol \, EDTA$$

$$M_1 V_1 = M_2 V_2$$

$$(M_{Ca \, solution})(V_{Ca \, solution, \, L})$$
$$= (M_{EDTA \, solution})(V_{EDTA \, solution})$$

Solve for $M_{EDTA \, solution}$

Titration of water sample with EDTA solution:

Trial	Dilution	Buret Start (ml)	Buret End (ml)	Volume Titrant (ml)	g Ca/ L	mg CaCO₃/ L
1						
2						
3						
					$\overline{X} =$	$\overline{X} =$
					$SD =$	$SD =$

Calcium content of water sample (g Ca/L and g CaCO₃/L):

$$mol \, calcium = mol \, EDTA$$

$$M_1 V_1 = M_2 V_2$$

$$(M_{Ca \, in \, sample})(V_{sample, \, L})$$
$$= (M_{EDTA \, solution})(V_{EDTA \, solution \, used \, in \, titration, \, L})$$

Solve for $M_{Ca \, in \, sample}$

$$M_{Ca \, in \, sample} \times 40.085 \, g \, Ca/mol = g \, Ca/L$$

$$(g \, Ca/L)(100.09 \, g \, CaCO_3/40.085 \, gCa) \times (1000 \, mg/g)$$
$$= mg \, CaCO_3/L$$

Questions

1. If a sample of water is thought to have a hardness of approximately 250 mg/L CaCO₃, what size sample (i.e., how many ml) would you use so that you would use approximately 10 ml of your EDTA solutions?
2. Why were you asked to prepare the CaCl₂ solution by using CaCO₃ and HCl rather than just weighing out CaCl₂?
3. In this EDTA titration method, would overshooting the endpoint in the titration cause an over- or underestimation of calcium in the sample? Explain your answer.

METHOD B: TEST STRIPS FOR WATER HARDNESS

Note

All information given is for AquaChek test strips, from Environmental Test Systems, Inc., a HACH Company, Elkhart, IN. Other similar test strips could be used. Any anion (e.g., magnesium, iron, copper) that will bind the EDTA may interfere with the AquaChek test. Very strong bases and acids also may interfere.

Principle of Method

The test strips have a paper, impregnated with chemicals, that is, adhered to polystyrene for ease of handling. The major chemicals in the paper matrix are Calmagite and EDTA, and minor chemicals are added to minimize reaction time, give long-term stability, and maximize color distinction between levels of water hardness. The strips are dipped into the water to test for total hardness caused by calcium and magnesium. The calcium displaces the magnesium bound to EDTA, and the released magnesium binds to Calmagite, causing the test strip to change color.

Chemicals

	CAS No.	Hazards
Calcium carbonate ($CaCO_3$)	471-34-1	Harmful
Calmagite	3147-14-6	
Ethylenediaminetetraacetic acid, disodium salt ($Na_2EDTA \cdot 2H_2O$)	60-00-4	Irritant
Hydrochloric acid, concentrated (HCl)	7647-01-0	Corrosive
Other proprietary chemicals in test strip		

Reagents

(**It is recommended that this solution be prepared by the laboratory assistant before class.)

- Calcium standard solution, 1.000 mg $CaCO_3$/ml** Prepare as described in Method A, using $CaCO_3$ and concentrated HCl.

Hazards, Precautions, and Waste Disposal

No precautions are needed in use of the test strip. Adhere to normal laboratory safety procedures. Wastes likely may be put down the drain using a water rinse, but follow good laboratory practices outlined by environmental health and safety protocols at your institution.

Supplies

- AquaChek® Test Strips (Environmental Test Systems, Inc., a HACH Company, Elkhart, IN.

1-800-548-4381. Contact the company about receiving a complementary package of test strips to use for teaching.)
- 2 Beakers, 100 ml

Procedure

(*Note*: Test the same tap water, tap distilled water, and standard calcium solution as used in Method A.)

1. Dip the test strip into a beaker filled with water or the standard calcium solution. Follow instructions on strip about how to read it, relating color to ppm $CaCO_3$.
2. Convert ppm $CaCO_3$ as determined with the test strips to mg $CaCO_3$/L and g Ca/L.

Data and Calculations

Sample	Trial (ppm $CaCO_3$)			Trial (mg $CaCO_3$/L)			Trial (g Ca/L)		
	1	2	3	1	2	3	1	2	3
Tap water									
Tap distilled water									
Standard Ca solution									

Question

1. Compare and discuss the accuracy and precision of the EDTA titration and test strip methods to measure calcium carbonate contents of the water samples and the calcium standard solution.

REFERENCES

Carpenter, C.E., and Hendricks, D.G. 2003. Mineral analysis. Ch. 12, in *Food Analysis*, 3rd ed. S.S. Nielsen (Ed.), Kluwer Academic, New York.

Clesceri, L.S., Greenberg, A.E., and Eaton, A.D. (Eds.). 1998. Method 2340 Hardness, *Standard Methods for the Examination of Water and Wastewater*, 20th ed., pp. 236–239. American Public Health Association, Washington, DC.

Sodium Determination Using Ion Selective Electrodes, Mohr Titration, and Test Strips

INTRODUCTION

Background

Sodium content of foods can be determined by various methods, including an ion selective electrode (ISE), the Mohr or Volhard titration procedure, or indicator test strips. These methods are official methods of analysis for numerous specific products. These methods all are faster and less expensive procedures than analysis by atomic absorption spectroscopy or inductively coupled plasma-atomic emission spectroscopy. This experiment allows one to compare sodium analysis of several food products by ISE, Mohr titration, and Quantab® chloride titrators.

Reading Assignment

Carpenter, C.E., and Hendricks, D.G. 2003. Mineral analysis. Ch. 12, in *Food Analysis*, 3rd ed. S.S. Nielsen (Ed.), Kluwer Academic, New York.

METHOD A: ION SELECTIVE ELECTRODES

Objective

Determine the sodium content of various foods with sodium and/or chloride ion selective electrodes.

Principle of Method

The principle of ISE is the same as for measuring pH, but by varying the composition of the glass in the sensing electrode, the electrode can be sensitive to sodium or chloride ions. Sensing and reference electrodes are immersed in a solution that contains the element of interest. The electrical potential that develops at the surface of the sensing electrode is measured by comparing the reference electrode with a fixed potential. The voltage between the sensing and reference electrodes relates to the activity of the reactive species. Activity (A) is related to concentration (C) by $A = \gamma C$, where γ is the activity coefficient, which is a function of ionic strength. By adjusting the ionic strength of all test samples and standards to a nearly constant (high) level, the Nernst equation can be used to relate electrode response to concentration of species being measured.

Chemicals

	CAS No.	Hazards
Ammonium chloride (NH$_4$Cl)	12125-02-9	Harmful
Ammonium hydroxide (NH$_4$OH)	1336-21-6	Corrosive, dangerous for environment
Nitric acid (HNO$_3$)	7697-37-2	Corrosive
Potassium nitrate (KNO$_3$)	7757-79-1	
Sodium chloride (NaCl)	7647-14-5	Irritant
Sodium nitrate (NaNO$_3$)	7631-99-4	Harmful, oxidizing

Reagents

Note: You can use a chloride and/or sodium ion selective electrode, with the appropriate associated solutions: electrode rinse solution, ionic strength adjustor, reference electrode fill solution, standard solution, electrode storage solution. Catalog numbers are given below for solutions from pHoenix Electrode Co., Houston TX.

- Electrode rinse solution
  For sodium electrode: Dilute 20 ml Ionic Strength Adjustor to 1 L with deionized distilled (dd) water. For chloride electrode: deionized distilled water
- Ionic strength adjuster (ISA)
  For sodium electrode: 4 M NH$_4$Cl, 4 M NH$_4$OH, Cat. # NAOISO1. For chloride electrode: 5 M NaNO$_3$, Cat. # CLOISO1.
- Nitric acid, 0.1 N
  Dilute 6.3 ml conc. HNO$_3$ to 1 L with dd water.
- Reference electrode fill solution
  For sodium electrode: 0.1 M NH$_4$Cl, Cat. # R001041. For chloride electrode: 10% KNO$_3$, Cat. # R001015.
- Standard solutions, 1000 ppm, sodium (Cat. # NAOASO2) and/or chloride (Cat. # CLOASO2)
  Use the 1000 ppm sodium or chloride solution to prepare 50 ml each of the following concentrations: 10, 20, 100, 500, and 1000 ppm sodium or chloride.

Hazards, Precautions, and Waste Disposal

Adhere to normal laboratory safety procedures. Wear gloves and safety glasses at all times. Ammonium hydroxide waste should be discarded as a hazardous waste. Other waste likely can be put down the drain using a water rinse, but follow good laboratory practices outlined by environmental health and safety protocols at your institution.

Supplies

- 15–30 Beakers, 250 ml (or sample cups to hold 100 ml)
- Food products: catsup, cottage cheese, potato chips, sports drink (e.g., Gatorade, white or clear)
- Graduated cylinder, 100 ml
- Magnetic stir bars
- Pipette bulb or pump
- 3 Spatulas
- 16–18 Volumetric flasks, 100 ml
- 2 Volumetric flasks, 50 ml
- Volumetric pipette, 2 ml
- 9 Volumetric pipettes, 5 ml
- Watch glass
- Weighing paper

Equipment

- Analytical balance
- Direct concentration readout ISE meter (i.e., suitable meter with millivolt accuracy to 0.1 mV)
- Heating plate with stirrer
- Magnetic stirrer
- Chloride electrode (e.g., pHoenix Electrode, Houston, TX, Chloride Ion Electrode, Cat. # CL01508-003B)
- Sodium electrode (e.g., pHoenix Electrode, Houston, TX, Sodium Ion Electrode, Cat. # NA71508-003B)

Procedure

(Replicate the preparation and analysis of standards and samples as specified by instructor.)

I. Sample Preparation (General Instructions)

1. Prepare samples by adding 5 g or 5 ml of sample (prehomogenized if necessary, and diluted if necessary) to a 100-ml volumetric flask. (See instructions specific for each type of food product below. Samples with high fat levels may require fat removal. Consult technical services of the company that manufacturers the ISE.) Add 2 ml ISA, then dilute to volume with dd water.
2. Prepare standards by adding 5 ml standard of proper dilution (e.g., 10, 20, 100, 500, 1000 ppm sodium or chloride) to a 100-ml volumetric flask. Add 2 ml ISA, then dilute to volume with dd water.

    Note: Sample/standard preparation calls for identical 1:20 dilution of each (i.e., 5 ml diluted to 100 ml). Therefore since samples

and standards are treated the same, no correction for this dilution needs to be made in calibration or calculation of results.

Specific Samples:

Sports drink: No dilution is required before a 5-ml sample is combined with the 2-ml ISA and dd water as described above.

Catsup: Accurately weigh ca. 1 g catsup into 50-ml volumetric flask, and dilute to volume with dd water. Mix well. Combine 5 ml of this diluted sample with 2-ml ISA and dd water as described above.

Cottage Cheese: Accurately weigh ca. 1 g of finely grated cheese into a 250-ml beaker containing a stir bar. Add 100 ml 0.1 N HNO_3. Cover beaker with a watch glass and boil gently for 20 min on stirrer/hot plate in a hood. Remove from hot plate and cool to room temperature in the hood. Pipette 5 ml of extract into 100-ml volumetric flask. Add 2 ml ISA and dilute to volume with dd water.

Potato chips: Accurately weigh ca. 5 g of potato chips into a 250-ml beaker. Crush chips with a glass stirring rod. Add 95 ml boiling dd water and stir. Filter water extract into a 100-ml volumetric flask, using a funnel with glass wool. Let cool to room temperature and dilute to volume.

II. Analysis

1. Condition sodium electrode as specified by the manufacturer.
2. Assemble, prepare, and check sodium and reference electrodes as described in electrode instruction manuals.
3. Connect electrodes to meter according to meter instruction manual.
4. For instruments with direct concentration readout capability, consult meter manual for correct direct measurement procedures.
5. Using pH meter set on mV scale, determine the potential (mV) of each standard solution (1, 10, 100, 500, 1000 ppm), starting with the most dilute standard. Use a uniform stirring rate, with a magnetic stir bar in each solution, sitting on a magnetic stir plate.
6. Rinse electrodes with electrode rinse solution between standards.
7. Measure samples and record the mV reading. As you rinse electrodes with electrode rinse solution between measurements, be careful not to get rinse solutions into the hole for outerfill solution in the reference electrode (or ensure that the hole is covered).

8. After use, store sodium electrode and reference electrode as specified by manufacturer.

Data and Calculations

1. Prepare a standard curve using 5-cycle semilog paper, with concentration plotted on the log axis. Plot actual concentration values on the log scale; not log values. Concentrations may be determined by reading directly off the standard curve, or using a calculated equation of the line. (Note: If the standard curve is really a curve and not a straight line, read directly off the curve rather than using an equation of the line.)

2. Use the standard curve and the mV readings for the samples to determine the sodium and/or chloride concentrations in ppm for the food samples as analyzed.

3. Convert the ppm sodium and/or chloride values for the food samples to mg/ml for the sports drink, catsup, cheese, and potato chips.

4. Taking into account the dilution of the samples, calculate the sodium and/or chloride content for catsup, cheese, and potato chips (in mg/g) (on a wet weight basis). Summarize the data and calculated results in one table. Show all sample calculations below each table.

5. Calculate sodium chloride content of each food, based on the chloride content.

6. Calculate the sodium content of each food, based on the sodium chloride content.

7. Compare the sodium chloride contents of the foods you analyzed to those reported in the U.S. Department of Agriculture (USDA) Nutrient Database for Standard Reference (http://www.nal.usda.gov/fnic/foodcomp/).

Question

1. If you used both a sodium and chloride ISE, which electrode worked better, concerning accuracy, precision, and time to response? Explain your answer, with appropriate justification.

METHOD B: MOHR TITRATION

Objective

Determine the sodium content of various foods using the Mohr titration method to measure chloride content.

Principle of Method

The Mohr titration is a direct titration method to quantitate chloride ions, to then calculate sodium ions. The chloride-containing sample solution is titrated with a standard solution of silver nitrate. After the silver from silver nitrate has complexed with all the available chloride in the sample, the silver reacts with chromate that has been added to the sample, to form an orange-colored solid, silver chromate. The volume of silver used to react with the chloride is used to calculate the sodium content of the sample.

Chemicals

	CAS No.	Hazards
Potassium chloride (KCl)	7447-40-7	Irritant
Potassium chromate (K_2CrO_4)	7789-00-6	Toxic, dangerous for environment
Silver nitrate ($AgNO_3$)	7761-88-8	Corrosive, dangerous for environment

Reagents

(**It is recommended that these solutions be prepared by laboratory assistant before class.)

- Potassium chloride
- Potassium chromate, 10% solution**
- Silver nitrate solution, ca. 0.1 N**
  Prepare approximately 400 ml of the ca. 0.1 M $AgNO_3$ (molecular weight (MW) 169.89) for each student or lab group. Students should accurately standardize the solution, as described in the Procedure.

Hazards, Precautions, and Waste Disposal

Wear gloves and safety glasses at all times, and use good lab technique. Potassium chromate may cause serious skin sensitivity reactions. Use of crystalline $AgNO_3$ or solutions of the silver salt can result in dark brown stains caused by photodecomposition of the salt to metallic silver. These stains are the result of poor technique on the part of the analyst, with spilled $AgNO_3$ causing discoloration of the floor. If you do spill this solution, immediately sponge up the excess solution and thoroughly rinse out the sponge at a sink. Then come back with the clean, rinsed sponge and mop up the area at least 3-4 times to remove all of the silver nitrate. Also, be sure to rinse all pipettes, burets, beakers, flasks, etc. to remove residual $AgNO_3$ when you are finished with this experiment. Otherwise these items also will stain, and drip stains are likely to appear on the floor. Potassium chromate and silver nitrate must be disposed of as a hazardous waste. Other waste likely can be put down the drain using a water rinse, but follow good laboratory practices outlined by environmental health and safety protocols at your institution.

Supplies

- 9 Beakers, 250 ml
- Brown bottle, 500 ml
- Buret, 25 ml
- 3 Erlenmeyer flasks, 125 ml
- 4 Erlenmeyer flasks, 250 ml
- Food products: cottage cheese (30 g), potato chips (15 g), sports drink (15 ml) (e.g., Gatorade, white or clear)
- Funnel
- Glass wool
- Graduated cylinder, 25 ml
- Magnetic stir bars (to fit 125 or 250 ml flasks)
- Pipette bulb or pump
- Spatulas
- Weighing paper and boats
- Volumetric pipette, 1 ml

Equipment

- Analytical balance
- Hot plate
- Magnetic stir plate

Procedure

(Instructions are given for analysis in triplicate.)

I. Standardization of ca. 0.1 M AgNO$_3$

1. Transfer 400 ml of the 0.1 M AgNO$_3$ solution to a brown bottle. This solution will be standardized, then used to titrate the food samples. Fill a buret with this AgNO$_3$ solution.
2. Prepare the primary standard (KCl, MW = 74.55) solution in triplicate. Accurately weigh to 4 decimal places about 100 mg KCl into three 125-ml Erlenmeyer flasks. Dissolve in dd water (about 25 ml), add 2–3 drops of K$_2$CrO$_4$ solution. CAUTION: potassium chromate may cause serious skin sensitivity reactions!)
3. Put a magnetic stir bar in each flask with the KCl solution, and place the beaker on a magnetic stir plate below the buret for titration. Using the AgNO$_3$ solution in the buret, titrate the KCl solutions to the appearance of the first permanent, pale, pink-orange color. (Note: you will first get a white precipitate, then green color, and then the pink-orange color.) This endpoint is due to formation of Ag$_2$CrO$_4$. The solution must be vigorously stirred during addition of the AgNO$_3$ solution to avoid erroneous results.
4. Record volume of AgNO$_3$.

5. Calculate and record molarity of AgNO$_3$.

$$\frac{\text{g KCl}}{(\text{ml AgNO}_3)} \times \frac{1 \text{ mol KCl}}{74.555 \text{ g}} \times \frac{1000 \text{ ml}}{1 \text{ L}}$$
$$= M \text{ of AgNO}_3/\text{L} = M \text{ AgNO}_3$$

6. Label bottle of AgNO$_3$ with your name and the molarity of the solution.

II. Determination of Chloride Content by Mohr Titration

Cottage Cheese

1. Accurately weigh 10 g of cottage cheese in triplicate into 250-ml beakers.
2. Add about 15 ml of warm dd water (50–55°C) to each beaker. Mix to a thin paste using a glass stirring rod or spatula. Add another ca. 25 ml dd water to each beaker until sample is dispersed.
3. Quantitatively transfer each solution to a 100-ml volumetric flask, rinsing beaker and magnetic stir bar with dd water several times. Dilute to volume with dd water.
4. Filter each solution through glass wool. Transfer 50 ml of each solution to 250-ml Erlenmeyer flasks.
5. Add 1 ml of potassium chromate indicator to each 50 ml of filtrate.
6. Titrate each solution with standardized ca. 0.1 M AgNO$_3$, to the first visible pale red-brown color that persists for 30 s. Record the volume of titrant used.

Potato Chips

1. Weigh accurately approximately 5 g of potato chips in duplicate into 250- ml beakers, then add 95 ml boiling dd water to each beaker.
2. Stir mixture vigorously for 30 s, wait 1 min, stir again for 30 s, then let cool to room temperature.
3. Filter each solution through glass wool. Transfer 50 ml of each solution to 250-ml Erlenmeyer flasks.
4. Add 1 ml of potassium chromate indicator to each 50 ml of filtrate.
5. Titrate each solution with standardized ca. 0.1 M AgNO$_3$, to the first visible pale red-brown color that persists for 30 s. Record the volume of titrant used.

Sports Drink (clear or white)

1. Pipette accurately 5 ml of sports drink in duplicate into 250-ml beakers, then add 95 ml boiling dd water to each beaker.
2. Stir mixture vigorously for 30 s, wait 1 min, stir again for 30 s.

3. Transfer 50 ml of each solution to 250-ml Erlen-meyer flasks.
4. Add 1 ml of potassium chromate indicator to each 50 ml of prepared sample.
5. Titrate each solution with standardized ca. 0.1 M AgNO$_3$, to the first visible pale red-brown color that persists for 30 s. Record the volume of titrant used.

Data and Calculations

1. Calculate the chloride content and the sodium chloride content of each replicated sample, then calculate the mean and standard deviation for each type of sample. Express the values in terms of percent, wt/vol, for the cottage cheese and potato chips, and percent, vol/vol, for the sports drink. Note that answers must be multiplied by the dilution factor.

$$\% \text{ chloride} = \frac{\text{ml of AgNO}_3}{\text{g (or ml) sample}} \times \frac{\text{mol AgNO}_3}{\text{L}}$$
$$\times \frac{35.5 \text{ g Cl}}{\text{mol NaCl}} \times \frac{1 \text{ L}}{1000 \text{ ml}} \times 100 \times \text{dilution factor}$$

$$\% \text{ sodium chloride (salt)} = \frac{\text{ml of AgNO}_3}{\text{g (or ml) sample}} \times \frac{\text{mol AgNO}_3}{\text{L}}$$
$$\times \frac{58.5 \text{ g}}{\text{mol NaCl}} \times \frac{1 \text{ L}}{1000 \text{ ml}} \times 100 \times \text{dilution factor}$$

Sample	Trial	Buret Start (ml)	Buret End (ml)	Vol. AgNO₃ (ml)	% Cl	% NaCl
Cottage cheese	1					
	2					
	3					
					$\overline{X} =$	
					SD $=$	
Potato chips	1					
	2					
	3					
					$\overline{X} =$	
					SD $=$	
Sports drink	1					
	2					
	3					
					$\overline{X} =$	
					SD $=$	

Questions

1. Show the calculations of how to prepare 400 ml of an approximately 0.1 M solution of AgNO$_3$ (MW = 169.89).

2. Would this Mohr titration procedure as described above work well to determine the salt content of grape juice or catsup? Why or why not?
3. How did this method differ from what would be done using a Volhard titration procedure? Include in your answer what additional reagents would be needed.
4. Would overshooting the endpoint result in an over- or underestimation of the salt content using the: (a) Mohr titration, (b) Volhard titration?

METHOD C: QUANTAB® TEST STRIPS

Objective

To measure the chloride content of foods using Quantab® Chloride Titrators, then calculate the sodium chloride content.

Principle of Method

Quantab® Chloride Titrators are thin, chemically inert plastic strips. These strips are laminated with an absorbent paper impregnated with silver nitrate and potassium dichromate, which together form brown silver dichromate. When the strip is placed in an aqueous solution that contains chlorine, the liquid rises up the strip by capillary action. The reaction of silver dichromate with chloride ions produces a white column of silver chloride in the strip. When the strip is completely saturated with the liquid, a moisture-sensitive signal across the top of the titrator turns dark blue to indicate the completion of the titration. The length of the white color change is proportional to the chloride concentration of the liquid being tested. The value on the numbered scale is read at the tip of the color change, and then is converted to percent salt using a calibration table.

Chemicals

	CAS No.	Hazards
Sodium chloride (NaCl)	7647-14-5	Irritant

Reagents

- Sodium chloride stock solution
  Accurately weigh 5.00 g of dried sodium chloride and quantitatively transfer to a 100-ml volumetric flask. Dilute to volume with dd water and mix thoroughly.
- Sodium chloride standard solutions
  Dilute 2 ml of the stock solution to 1000 ml with dd water in a volumetric flask to create a 0.010% sodium chloride solution to use as a standard

solution with the low range Quantab® Chloride Titrators.

Dilute 5 ml of the stock solution to 100 ml with dd water in a volumetric flask to create a 0.25% sodium chloride solution to use as a standard solution with the high range Quantab® Chloride Titrators.

Supplies

- 5 Beakers, 200 ml
- Filter paper (when folded as a cone, should fit into a 200-ml beaker)
- Glass wool
- Glass stirring rod
- Graduated cylinder, 100 ml
- Quantab® Chloride Titrators, range: 0.05 to 1.0% (high range) and 0.005 to 0.1% (low range) expressed as NaCl. (Environmental Test Systems, Inc., Elkhart, IN, 1-800-548-4381). Contact the company about receiving a complimentary package of Quantab® Chloride Titrators to use for a teaching laboratory.
- Sports drink, 10 ml (i.e., same one used in Methods A and B)
- Volumetric flask, 100 ml

Equipment

- Hot plate
- Top loading balance

Procedure

(Instructions are given for analysis in triplicate.)

I. Standard Solutions of Sodium Chloride

1. Transfer 50 ml of the 0.25% standard sodium chloride solution to a 200- ml beaker.
2. Fold a piece of filter paper into a cone-shaped cup and place it point end down into the beaker. This will allow liquid from the beaker to seep through the filter paper at the pointed end.
3. Using the 0.25% sodium chloride standard solution, place the lower end of the High Range Quantab® Strip (0.05 to 1.0%) into the filtrate within the pointed end of the filter paper cone, being sure not to submerge the titrator more than 1.0 in.
4. Thirty sec after the moisture-sensitive signal string at the top of the titrator turns dark blue or a light brown, record the Quantab® reading at the tip of the yellow-white peak, to the nearest 0.1 units on the titrator scale.

5. Using the calibration chart included with the Quantab® package, convert the Quantab® reading to percent sodium chloride (NaCl) and to ppm chloride (Cl⁻). Note that each lot of Quantab® has been individually calibrated. Be sure to use the correct calibration chart, (i.e., the control number on the product being used must match the control number on the bottle).
6. Repeat Steps 1–5 above using the 0.01% sodium chloride standard solution with the Low Range Quantab® Strip.

II. Cottage Cheese

1. Weigh accurately approximately 5 g of cottage cheese into a 200-ml beaker, then add 95 ml boiling dd water.
2. Stir mixture vigorously for 30 s, wait 1 min, stir again for 30 s, then let cool to room temperature.
3. Fold a piece of filter paper into a cone-shaped cup and place it point end down into the beaker. This will allow liquid from the beaker to seep through the filter paper at the pointed end.
4. Testing with both the Low Range and the High Range Quantab® test strips, place the lower end of the Quantab® into the filtrate within the pointed end of the filter paper cone, being sure not to submerge the titrator more than 2.5 cm.
5. Thirty sec after the moisture-sensitive signal string at the top of the titrator turns dark blue or a light brown, record the Quantab® reading at the tip of the yellow-white peak, to the nearest 0.1 units on the titrator scale.
6. Using the calibration chart included with the Quantab® package, convert the Quantab® reading to percent sodium chloride (NaCl) and to ppm chloride (Cl⁻). Note that each lot of Quantab® has been individually calibrated. Be sure to use the correct calibration chart (i.e., the control number on the product being used must match the control number on the bottle).
7. Multiply the result by the dilution factor 20 to obtain the actual salt concentration in the sample.

III. Potato Chips

1. Weigh accurately approximately 5 g of potato chips into a 200-ml beaker. Crush chips with a glass stirring rod. Add 95 ml boiling dd water and stir.
2. Filter water extract into a 100-ml volumetric flask, using a funnel with glass wool. Let cool to room temperature and dilute to volume. Transfer to a 200-ml beaker.
3. Follow Steps 3–7 from procedure for cottage cheese.

IV. Catsup

1. Weigh accurately approximately 5 g of catsup into a 200-ml beaker. Add 95 ml boiling dd water and stir.
2. Filter water extract into a 100-ml volumetric flask. Let cool to room temperature and dilute to volume. Transfer to a 200-ml beaker.
3. Follow Steps II.3–7 from procedure for cottage cheese.

V. Sports Drink

1. Weigh accurately approximately 5 ml of sports drink into a 200-ml beaker. Add 95 ml boiling dd water and stir.
2. Follow Steps II.3–7 from procedure for cottage cheese.

Data and Calculations

	From Calibration Chart				Corrected for Dilution Factor			
	% NaCl		ppm Cl		% NaCl		ppm Cl	
Trial	LR	HR	LR	HR	LR	HR	LR	HR
Catsup								
1								
2								
3								
					$\overline{X} =$ SD =	$\overline{X} =$ SD =	$\overline{X} =$ SD =	$\overline{X} =$ SD =
Cottage cheese								
1								
2								
3								
					$\overline{X} =$ SD =	$\overline{X} =$ SD =	$\overline{X} =$ SD =	$\overline{X} =$ SD =
Potato chips								
1								
2								
3								
					$\overline{X} =$ SD =	$\overline{X} =$ SD =	$\overline{X} =$ SD =	$\overline{X} =$ SD =
Sports drink								
1								
2								
3								
					$\overline{X} =$ SD =	$\overline{X} =$ SD =	$\overline{X} =$ SD =	$\overline{X} =$ SD =

SUMMARY OF RESULTS

Summarize in a table the sodium chloride content (mean and standard deviation) of the various food products as determined by the three methods described in this experiment. Include in the table the sodium chloride contents of the foods from the nutrition label and those published in the USDA Nutrient Database for Standard Reference (web address: http://www.nal.usda.gov/fnic/foodcomp/).

Sodium chloride content (%) of foods by various methods:

Food Product	Ion selective Electrode	Mohr Titration	Quantab® Titration	Nutrition Label	USDA Data
Catsup	$\overline{X} =$ SD =				
Cottage cheese	$\overline{X} =$ SD =				
Potato chips	$\overline{X} =$ SD =				
Sports drink	$\overline{X} =$ SD =				

QUESTIONS

1. Based on the results and characteristics of the methods, discuss the relative advantages and disadvantages of each method of analysis for these applications.
2. Comparing your results to data from the nutrition label and USDA Nutrient Database, what factors might explain any differences observed?

ACKNOWLEDGMENTS

Phoenix Electrode, Houston, TX, is acknowledged for its contribution of the sodium and chloride ion selective electrodes, and related supplies, for use in developing a section of this laboratory exercise. Environmental Test Systems, Inc., a HACH Company, Elkhart, IN, is acknowledged for its contributing the Quantab® Chloride Titrators for use in developing a section of this laboratory exercise.

REFERENCES

AOAC International. 2000. Method 941.18, Standard Solution of Silver Nitrate; Method 983.14, Chloride (total) in cheese, *Official Methods of Analysis*, 17th ed. AOAC International, Gaithersburg, MD.

AOAC International. 2000. Method 971.19, Salt (Chlorine as Sodium Chloride) in Meat, Fish, and Cheese, Indicating Strip Method, *Official Methods of Analysis*, 17th ed. (refers to 15th ed., 1990). AOAC International, Gaithersburg, MD.

AOAC International. 2000. Method 976.25, Sodium in Foods for Special Dietary Use, Ion Selective Electrode Method, *Official Methods of Analysis*, 17th ed. AOAC International, Gaithersburg, MD.

Carpenter, C.E., and Hendricks, D.G. 2003. Mineral analysis. Ch. 12, in *Food Analysis*, 3rd ed. S.S. Nielsen (Ed.), Kluwer Academic, New York.

Environmental Test Systems. 2001. Quantab® Technical Bulletin. Chloride Analysis for Cottage Cheese. Environmental Test Systems, Elkhart, IN.

Phoenix Electrode, Houston, TX. Product literature.

Wehr, H.M., and Frank, J.F. (Eds.) 2002. Part 15.053 Chloride (Salt), *Standard Methods for the Examination of Dairy Products*, 17th ed. American Public Health Association, Washington, DC.

10
chapter

Sodium and Potassium Determinations by Atomic Absorption Spectroscopy and Inductively Coupled Plasma-Atomic Emission Spectroscopy

INTRODUCTION

Background

The concentration of specific minerals in foods can be determined by a variety of methods. The purpose of this lab is to acquaint you with the use of atomic absorption spectroscopy (AAS) and atomic emission spectroscopy (AES) for that use. The specific type of AES will be inductively coupled plasma-atomic emission spectroscopy (ICP-AES).

In recent years, some instrument manufacturers have started calling AES instruments "optical emission spectrometers" (OES) because they measure light emitted when excited atoms return to the ground state. In this experiment, the term AES will be used rather than OES, but the two terms are virtually interchangeable.

This experiment specifies the preparation of standards and samples for determining the sodium (Na) and potassium (K) contents by AAS and ICP-AES. Samples suggested for analysis include two solid food products that requires wet and/or dry ashing prior to analysis (catsup and potato chips) and one liquid food product that does not (a clear sports drink, or a clear fruit juice).

Procedures are described for both wet ashing and dry ashing of the solid samples, so experience can be gained with both types of ashing and so results can be compared between the two methods of ashing. Sodium results from this experiment can be compared to sodium results from analysis of the same products in the experiment that uses the more rapid methods of analysis of ion selective electrodes, the Mohr titration, and Quantab® test strips.

The limit of detection for sodium is 0.3 parts per billion (ppb) for flame AAS, 3 ppb by radial ICP-AES, and 0.5 ppb by axial ICP-AES. The limit of detection for potassium is 3 ppb for flame AAS, 0.2–20 ppb (depending on the model) by radial ICP-AES, and 1 ppb by axial ICP-AES. Other comparative characteristics of AAS and ICP-AES are described in Chapter 25 of Nielsen, *Food Analysis*.

Reading Assignment

Harbers, L.H., and S.S. Nielsen. 2003. Ash analysis. Ch. 7, in *Food Analysis*, 3rd ed. S.S. Nielsen (Ed.), Kluwer Academic, New York.

Miller, D.D., and M.A. Rutzke. 2003. Atomic absorption and emission spectroscopy. Ch. 25, in *Food Analysis*, 3rd ed. S.S. Nielsen (Ed.), Kluwer Academic, New York.

Note

If there is no access to an ICP-AES, a simple AES unit can be used, likely with the same standard solutions and samples prepared as described below.

Objective

The objective of this experiment is to determine the sodium and potassium contents of food products using atomic absorption spectroscopy and inductively coupled plasma-atomic emission spectroscopy.

Principle of Method

Atomic absorption is based on atoms *absorbing* energy, once heat energy from a flame has converted molecules to atoms. By absorbing the energy, atoms go from ground state to an excited state. The energy absorbed is of a specific wavelength from a hollow cathode lamp. One measures absorption as the difference between the amount of energy emitted by the hollow cathode lamp and that reaching the detector. Absorption is linearly related to concentration.

Atomic emission is based on atoms *emitting* energy, once heat energy from a flame has converted molecules to atoms then raised the atoms from ground state to an excited state. The atoms emit energy of a specific wavelength as they drop from an excited state back down to ground state. One measures the amount of emitted energy of a wavelength specific for the element of interest. Emission is linearly related to concentration.

Chemicals

	CAS No.	Hazards
Hydrochloric acid (HCl)	7647-01-1	Corrosive
Hydrogen peroxide, 30% (H_2O_2)	7722-84-1	Corrosive
Lanthanum chloride ($LaCl_3$)	10025-84-0	Irritant
Nitric acid (HNO_3)	7697-37-2	Corrosive
Potassium chloride (KCl) (for K std. solution)	7447-40-7	Irritant
Sodium chloride (NaCl) (for Na std. solution)	7647-14-5	Irritant

Reagents

(**It is recommended that these solutions be prepared by laboratory assistant before class.)

- Potassium and sodium standard solutions, 1000 ppm**
  Used to prepare 100 ml solutions of each of the concentrations listed in Table 10-1. Each standard solution must contain 10 ml conc. HCl/100 ml final volume.

Hazards, Precautions, and Waste Disposal

Adhere to normal laboratory safety procedures. Wear safety glasses and gloves during sample preparation. Use acids in a hood.

10-1
table

Concentrations (ppm) of Na and K Standard Solutions for AAS and ICP-AES

AAS		ICP-AES	
Na	K	Na	K
0.20	0.10	50	50
0.40	0.50	100	100
0.60	1.00	200	200
0.80	1.50	300	300
1.00	2.00	400	400

Supplies

(Used by students)

- 2 Crucibles, previously cleaned and heated at 550°C in a muffle furnace for 18 hr (for dry ashing)
- Dessicator, with dry dessicant
- Digestion tubes (for wet ashing; size to fit digestion block)
- Filter paper, ashless
- Funnels, small (to filter samples)
- Plastic bottles, with lids, to hold 50 ml (or plastic sample tubes with lids, to hold 50 ml, to fit autosampler, if one is available)
- 8 Volumetric flasks, 25 ml
- 4 Volumetric flasks, 50 ml
- Volumetric flask, 100 ml
- Volumetric pipettes, 2 ml, 4 ml, 5 ml, 10 ml (2)
- Weigh boats/paper

Equipment

- Analytical balance
- Atomic absorption spectroscopy unit
- Digestion block (for wet ashing; set to 175°C)
- Inductively coupled plasma-atomic absorption spectroscopy unit (or simple atomic absorption spectroscopy unit)
- Muffle furnace (for dry ashing; set to 550°C)
- Water bath, heated to boil water (for dry ashing)

PROCEDURE

Sample Preparation: Liquid Samples

1. Put an appropriate volume of liquid sample in a 100-ml volumetric flask. For a sports drink, use 0.2 ml for both Na and K analysis by AAS. Use 50 ml for Na analysis and 80 ml for K analysis by ICP-AES.
2. Add 10 ml conc. HCl.
3. Add deionized distilled (dd) water to volume.
4. Shake well. (If there is any particulate matter present, the sample will need to be filtered through ashless filter paper.)
5. Make appropriate dilution and analyze (to sample for AAS, add LaCl$_3$ to final conc. of 0.1%).

Liquid Blank:

Prepare a liquid blank sample to be assayed, following the sample preparation procedure but excluding the sample.

Sample Preparation: Solid Samples

I. Wet Ashing

Note: Digestion procedure described is a wet digestion with nitric acid and hydrogen peroxide. Other types of digestion can be used instead.

1. Label one digestion tube per sample plus one tube for the reagent blank (control).
2. Accurately weigh out 300–400 mg of each sample and place in a digestion tube. Prepare samples in duplicate or triplicate.
3. Pipette 5 ml concentrated nitric acid into each tube, washing the sides of the tube as you add the acid.
4. Set tubes with samples and reagent blank in digestion block. Turn on the digestion block and set to 175°C to start the predigestion.
5. Swirl the samples gently once or twice during the nitric acid predigestion, using tongs and protective gloves.
6. Remove tubes from digestion block when brown gas starts to elute (or when solution begins to steam, if there is no brown gas) and set in cooling rack. Turn off digestion block.
7. Let samples cool for at least 30 min. (Samples can be stored at this point for up to 24 hr.)
8. Add 4 ml of 30% hydrogen peroxide to each tube, doing only a few tubes at one time. Gently swirl tubes. Put tubes back in digestion block. Turn on digestion block still set to 175°C.
9. Watch tubes closely for the start of the reaction, indicated by the appearance of rapidly rolling bubbles. As soon as the reaction starts, remove the tubes from the block and let the reaction continue in the cooling rack. (Caution: Some sample types will have a vigorous reaction, and for some the sample is lifted to the top of the tube, with risk of boiling over.)
10. Repeat Steps 8 and 9 for all samples and the reagent blank.
11. Put all tubes in the digestion block, and leave until ca. 1–1.5 ml remains, then remove

10-2 table	**Dilution of Samples for Na and K analysis by AAS and ICP-AES, Using Wet[1] or Dry Ashing[2]**			
	Na		K	
Sample	AAS	ICP-AES	AAS	ICP-AES
Catsup				
Wet ashing	Ashed sample diluted to 25 ml, then 0.2 ml diluted to 100 ml	Ashed sample diluted to 25 ml	Ashed sample diluted to 25 ml, then 0.4 ml diluted to 100 ml	Ashed sample diluted to 10 ml
Dry ashing	Ashed sample diluted to 25 ml, then 0.2 ml diluted to 100 ml	Ashed sample diluted to 50 ml	Ashed sample diluted to 25 ml, then 0.2 ml diluted to 100 ml	Ashed sample diluted to 25 ml
Cottage cheese				
Wet ashing	Ashed sample diluted to 25 ml, then 0.5 ml diluted to 100 ml	Ashed sample diluted to 10 ml	Ashed sample diluted to 25 ml, then 0.7 ml diluted to 100 ml	Ashed sample diluted to 5 ml
Dry ashing	Ashed sample diluted to 25 ml, then 0.2 ml diluted to 100 ml	Ashed sample diluted to 25 ml	Ashed sample diluted to 25 ml, then 0.5 ml diluted to 100 ml	Ashed sample diluted to 25 ml
Potato chips				
Wet ashing	Ashed sample diluted to 25 ml, then 0.2 ml diluted to 100 ml	Ashed sample diluted to 10 ml	Ashed sample diluted to 25 ml, then 0.2 ml diluted to 100 ml	Ashed sample diluted to 25 ml
Dry ashing	Ashed sample diluted to 25 ml, then 0.2 ml diluted to 100 ml	Ashed sample diluted to 25 ml	Ashed sample diluted to 50 ml, then 0.1 ml diluted to 100 ml	Ashed sample diluted to 50 ml

[1] For wet ashing, use ca. 300–400 mg sample.
[2] For dry ashing, use ca. 1 g sample, dry matter (calculate based on moisture content).

each tube from the digestion block. Check the tubes every 10–15 min during this digestion. (If the tubes are left on the digestion block too long and they become dry, remove, cool, and *carefully* add ca. 2 ml concentrated nitric acid and continue heating.) Turn off digestion block when all tubes have been digested and removed.

12. Make appropriate dilution of samples with dd water in a volumetric flask as indicated in Table 10-2. (To sample for AAS, add $LaCl_3$ to final conc. of 0.1%.)
13. If necessary, filter samples using Whatman hardened ashless #540 filter paper into container appropriate for analysis by AAS or ICP-AES.

II. Dry Ashing

1. Accurately weigh out blended or ground ca. l-g sample dry matter into crucible (i.e., take moisture content into account, so you have ca. 1 g dry product).
2. Pre-dry sample over boiling water bath.

3. Complete drying of sample in vacuum oven at 100°C, 26 in. Hg, for 16 hrs.
4. Dry ash sample for 18 hr. at 500°C, then let cool in dessicator.
5. Dissolve ash in 10 ml HCl solution (1:1, $HCl:H_2O$).
6. Make appropriate dilution of samples with dd water in a volumetric flask as indicated in Table 10-2. (To sample for AAS, add $LaCl_3$ to final conc. of 0.1%.)
7. If necessary, filter samples using Whatman hardened ashless #540 filter paper into container appropriate for analysis by AAS or ICP-AES.

Analysis

Follow manufacturer's instructions for startup, use, and shutdown of the AAS and ICP-AES. Take appropriate caution with the acetylene and flame in using AAS, and the liquid or gas argon and the plasma in using the ICP-AES. Analyze standards, reagent blanks, and samples.

DATA AND CALCULATIONS

Note: Due to the nature of the differences between printouts for various ICP-AES manufacturers, the ICP operator should assist with interpretation of ICP-AES results. As specified under data handling instruction below, if ICP-AES emission data are available for standards, they should be recorded and plotted, for comparison to AAS standard curves. If ICP-AES emission data are available for samples, they should be converted to concentration data in ppm using the appropriate standard curve. If ICP-AES emission data are not available, report concentration in ppm.

Do all calculations for each duplicate sample individually, before determining a mean value on the final answer.

Standard Curve Data

Potassium Standard Curves				Sodium Standard Curves			
AAS		ICP-AES		AAS		ICP-AES	
ppm	Absorption	ppm	Emission	ppm	Absorption	ppm	Emission
50		1		50		1	
100		5		100		5	
200		10		200		10	
300		20		300		20	

Sample Data

Atomic Absorption Spectroscopy

Sample	Trial	Sample Size (g or ml)	Absorption	ppm	Dilution	Diluted (mg/ml or g/g)	Original (mg/ml or mg/g)
Liquid blank	1				—	—	—
	2				—	—	—
Sports drink	1						
	2						
Solid blank	1				—	—	—
	2				—	—	—
Catsup	1						
	2						
Cottage chesse	1						
	2						
Potato chips	1						
	2						

Inductively Coupled Plasma-Atomic Emission Spectroscopy

Sample	Trial	Sample Size (g or ml)	Emission	ppm	Dilution	Diluted (mg/ml or g/g)	Original (mg/ml or mg/g)
Liquid blank	1				—	—	—
	2				—	—	—
Sports drink	1						
	2						
Solid blank	1				—	—	—
	2				—	—	—
Catsup	1						
	2						
Potato chips	1						
	2						

Data Handling

1. Prepare standard curves for sodium and potassium as measured by AAS.
2. Use the standard curves from AAS and the absorption readings of the samples to determine the concentrations in ppm of sodium and potassium for the food samples as analyzed (i.e., ashed and/or diluted).
   Note: For the AAS samples, you need to subtract the liquid blank absorbance from the sports drink sample values, and the solid blank absorbance from the catsup and potato chip sample values.
3. Prepare standard curves for sodium and potassium as measured by ICP-AES (if emission data are available).
4. Use the standard curves from ICP-AES and the emission readings of the samples to determine the concentrations in ppm of sodium and potassium for the food samples as analyzed (i.e., ashed and/or diluted). If emission data are not available for the samples, record the concentrations in ppm of sodium and potassium for the food samples as analyzed (i.e., ashed and/or diluted).
5. Convert the AAS and ICP-AES values for samples in ppm to mg/ml for the sports drink and to mg/g for the catsup and potato chips.
6. Calculate the sodium and potassium contents by AAS and ICP-AES for the original samples of sports drink (in mg/ml), catsup (in mg/g), and potato chips (mg/g) (on a wet weight basis). Summarize the data and calculated results in one

table for AAS and ICP-AES. Show examples of all calculations below the table.

QUESTIONS

1. Compare the sodium and potassium values for catsup and potato chips to those reported in the U.S. Department of Agriculture Nutrient Database for Standard Reference (http://www.nal.usda.gov/fnic/foodcomp). Which method of analysis gives a value closer to that reported in the database for sodium? For potassium?
2. Describe how you would prepare the Na and K standard solutions for AES, using the 1000 ppm solutions of each, which are commercially available. If possible, all solutions for points in the standard curve should be made using different volumes of the same stock solution. Do not use volumes of less than 0.2 ml. Make all standards to the same volume of 100 ml. Note that each standard solution must contain 10 ml conc. HCl/100 ml final volume, as described under Reagents.
3. Describe how you would prepare a 1000-ppm Na solution, starting with commercially available solid NaCl.

REFERENCES

Harbers, L.H., and S.S. Nielsen. 2003. Ash analysis. Ch. 7, in *Food Analysis*, 3rd ed. S.S. Nielsen (Ed.), Kluwer Academic, New York.

Miller, D.D., and M.A. Rutzke. 2003. Atomic absorption and emission spectroscopy. Ch. 25, in *Food Analysis*, 3rd ed. S.S. Nielsen (Ed.), Kluwer Academic, New York.

Standard Solutions and Titratable Acidity

INTRODUCTION

Background

Many types of chemical analyses are made by a method in which a constituent is titrated, using a solution of known strength, to an endpoint with an indicator. Such a solution is referred to as a standard solution. From the volume and concentration of standard solution used in the titration, and knowing the sample size, the concentration of the constituent in the sample can be calculated.

The assay for titratable acidity is a volumetric method that uses a standard solution and, most commonly, the indicator phenolphthalein. A standard solution of sodium hydroxide reacts in the titration with organic acids present in the sample. The normality of the sodium hydroxide solution, the volume used, and the volume of the test sample are used to calculate titratable acidity, expressing it in terms of the predominant acid present in the sample. A standard acid such as potassium acid phthalate can be used to determine the exact normality of the standard sodium hydroxide used in the titration.

The phenolphthalein endpoint in the assay for titratable acidity is pH 8.2, where there is a significant color change from clear to pink. When colored solutions obscure the pink endpoint, a potentiometric method is commonly used. A pH meter is used to titrate such a sample to pH 8.2.

Reading Assignment

Sadler, G.D., Murphy, P.A. 2003. pH and titratable acidity. Ch. 13, in *Food Analysis*, 3rd ed. S.S. Nielsen, Ed., Kluwer Academic, New York.

Notes

1. Carbon dioxide (CO_2) acts as an interfering substance in determining titratable acidity, by the following reactions:

$$H_2O + CO_2 \leftrightarrow H_2CO_3 \text{ (carbonate)}$$

$$H_2CO_3 \leftrightarrow H^+ + HCO_3^- \text{ (bicarbonate)}$$

$$HCO_3^- \leftrightarrow H^+ + CO_3^{-2}$$

In these reactions, buffering compounds and hydrogen ions are generated. Therefore, CO_2-free water is prepared and used for standardizing acids and base and for determining titratable acidity. An ascarite trap is attached to bottles of CO_2-free water, so that as air enters the bottle when water is siphoned out, the CO_2 is removed from the air.

2. Ascarite is a silica base coated with NaOH, and it removes CO_2 from the air by the following reaction:

$$2NaOH + CO_2 \rightarrow Na_2CO_3 + H_2O$$

METHOD A: PREPARATION AND STANDARDIZATION OF BASE AND ACID SOLUTIONS

Objective

Prepare and standardize solutions of sodium hydroxide and hydrochloric acid.

Principle of Method

A standard acid can be used to determine the exact normality of a standard base, and vice versa.

Chemicals

	CAS No.	Hazards
Ascarite	81133-20-2	Corrosive
Ethanol (CH_3CH_2OH)	64-17-5	Highly flammable
Hydrochloric acid (HCl)	7647-01-0	Corrosive
Phenolphthalein	77-09-8	Irritant
Potassium acid phthalate ($HOOCC_6H_4COOK$)	877-24-7	Irritant
Sodium hydroxide (NaOH)	1310-73-2	Corrosive

Reagents

(**It is recommended that these solutions be prepared by the laboratory assistant before class.)
(*Note*: Preparation of NaOH and HCl solutions is described under Procedure.)

- Ascarite trap**
  Put the ascarite in a syringe that is attached to the flask of CO_2-free water (see note about CO_2-free water).
- Carbon dioxide-free water
  Prepare 1.5 L of CO_2-free water (per person or group) by boiling deionized distilled (dd) water for 15 min in a 2-L Erlenmeyer flask. After boiling, stopper the flask with a rubber stopper through which is inserted a tube attached to an ascarite trap. Allow the water to cool with ascarite protection.
- Ethanol, 100 ml
- Hydrochloric acid, concentrated
- Phenolphthalein indicator solution, 1%**
  Dissolve 1.0 g in 100 ml ethanol. Put in bottle with eyedropper.

- Potassium acid phthalate (KHP)**
  3–4 g, dried in an oven at 120°C for 2 hr, cooled
  and stored in a closed bottle inside a desiccator
  until use
- Sodium hydroxide, pellets

Hazards, Precautions, and Waste Disposal

Use appropriate precautions in handling concentrated
acid and base. Otherwise, adhere to normal laboratory
safety procedures. Wear gloves and safety glasses at all
times. Waste likely may be put down the drain using a
water rinse, but follow good laboratory practices outlined by environmental health and safety protocols at
your institution.

Supplies

(Used by students)

- Beaker, 50 ml (for waste NaOH from buret)
- Beaker, 100 ml
- Buret, 25 or 50 ml
- 5 Erlenmeyer flasks, 250 ml
- Funnel, small, to fit top of 25 or 50 ml buret
- Glass stirring rod
- Glass storage bottle, 100 ml
- Graduated cylinder, 50 ml
- Graduated cylinder, 1 L
- Measuring pipette, 1 ml
- Measuring pipette, 10 ml
- Parafilm®
- Pipette bulb or pump
- Plastic bottle, with lid, 50 or 100 ml
- Plastic bottle, 1 L
- Spatula
- Squirt bottle, with dd water
- Volumetric flask, 50 ml
- Volumetric flask, 100 ml
- Weighing paper/boat
- White piece of paper

Equipment

- Analytical balance
- Hot plate
- Forced draft oven (heated to 120°C)

Calculations Required before Lab

1. Calculate how much NaOH to use to prepare
   50 ml of 25% NaOH (wt/vol) in water (see
   Preface for definition of percent solutions).
2. Calculate how much concentrated HCl to use
   to prepare 100 ml of ca. 0.1 N HCl in water
   (concentrated HCl = 12.1 N).

Procedure

1. Prepare 25% (wt/vol) NaOH solution: Prepare
   50 ml of 25% NaOH (wt/vol) in dd water. To do
   this, weigh out the appropriate amount of NaOH
   and place it in a 100-ml beaker. While adding
   about 40 ml of dd water, stir the NaOH pellets
   with a glass stirring rod. Continue stirring until
   all pellets are dissolved. Quantitatively transfer
   the NaOH solution into a 50-ml volumetric flask.
   Dilute to volume with dd water. The solution
   must be cooled to room temperature before final
   preparation. Store this solution in a plastic bottle
   and label appropriately.
2. Prepare ca. 0.1 N HCl solution: Prepare 100 ml of
   ca. 0.1 N HCl using concentrated HCl (12.1 N)
   and dd water. (Note: Do not use a mechanical
   pipettor to prepare this, since the acid can easily get into the shaft of the pipettor and cause
   damage.) To prepare this solution, place a small
   amount of dd water in a 100-ml volumetric flask,
   pipette in the appropriate amount of concentrated HCl, then dilute to volume with dd water.
   Mix well, and transfer into a glass bottle, seal
   bottle, and label appropriately.
3. Prepare ca. 0.1 N NaOH solution: Transfer 750 ml
   CO_2-free water to a 1-liter plastic storage bottle. Add ca. 12.0 ml of well-mixed 25% (wt/vol)
   NaOH solution prepared in Step 1. Mix thoroughly. This will give an approximately 0.1 N
   solution. Fill the buret with this solution using
   a funnel. Discard the first volume of the buret,
   then refill the buret with the NaOH solution.
4. Standardize ca. 0.1 N NaOH solution: Accurately weigh about 0.8 g of dried potassium
   acid phthalate (KHP) into each of three 250-
   ml Erlenmeyer flasks. Record the exact weights.
   Add ca. 50 ml of cool CO_2-free water to each
   flask. Seal the flasks with Parafilm® and swirl
   gently until the sample is dissolved. Add 3 drops
   of phenolphthalein indicator and titrate, against
   a white background, with the NaOH solution
   being standardized. Record the beginning and
   ending volume on the buret. Titration should
   proceed to the faintest tinge of pink that persists
   for 15 s after swirling. The color will fade with
   time. Record the total volume of NaOH used
   to titrate each sample. Data from this part will
   be used to calculate the mean normality of the
   diluted NaOH solution.
5. Standardize ca. 0.1 N HCl solution: Devise
   a scheme to standardize (i.e. determine the
   exact N) the ca. 0.1 N HCl solution that you prepared in Step 2. Remember that you have your
   standardized NaOH to use. Do analyses in at
   least duplicate. Record the volumes used.

Data and Calculations

Step 4

Using the weight of KHP and the volume of NaOH titrated in Step 4, calculate the normality of the diluted NaOH solution as determined by each titration, then calculate the mean normality (molecular weight (MW) of potassium acid phthalate = 204.228). The range of triplicate determinations for normality should be less than 0.2% with good technique.

Trial	Weight of KHP (g)	Buret Start (ml)	Buret End (ml)	Vol. NaOH Titrated (ml)	N NaOH
1					
2					
3					
					$\overline{X} =$
					SD =

Sample calculation:

Weight of KHP = 0.8115 g

MW of KHP = 204.228 g/mol

Volume of ca. 0.10 N NaOH used in
   titration = 39 ml

Mol KHP = 0.8115 g/204.228 g/mol
   = 0.003974 mol

Mol KHP = Mol NaOH
   = 0.003974 mol = N NaOH × L NaOH

0.003978 mol NaOH/0.039 L NaOH = 0.1019 N

Step 5

With the volumes of HCl and NaOH used in Step 5, calculate the exact normality of the HCl solution as determined by each titration, then calculate the mean normality.

Trial	Vol. HCl (ml)	Vol. NaOH (ml)	N HCl
1			
2			
			$\overline{X} =$

Questions

1. What does 25% NaOH (wt/vol) mean? How would you prepare 500 ml of a 25% NaOH (wt/vol) solution?
2. Describe how you prepared the 100 ml of ca. 0.1 N HCl. Show your calculations.

3. If you had not been told to use 12 ml of 25% NaOH (wt/vol) to make 0.75 L of ca. 0.1 N NaOH, how could you have determined this was the appropriate amount? Show all calculations.
4. Describe in detail how you standardized your ca. 0.1 N HCl solution.

METHOD B: TITRATABLE ACIDITY AND pH

Objective

Determine the titratable acidity and pH of food samples.

Principle of Method

The volume of a standard base used to titrate the organic acids in foods to a phenolphthalein endpoint can be used to determine the titratable acidity.

Chemicals

	CAS No.	Hazards
Ascarite	81133-20-2	Corrosive
Ethanol (CH_3CH_2OH)	64-17-5	Highly flammable
Phenolphthalein	77-09-8	Irritant
Sodium hydroxide (NaOH)	1310-73-2	Corrosive

Reagents

(**It is recommended that these items/solutions be prepared by the laboratory assistant before class.)

- Ascarite trap**
  Put the ascarite in a syringe that is attached to the flask of CO_2-free water.
- Carbon dioxide-free water**
  Prepared and stored as described in Method A.
- Phenolphthalein indicator solution, 1%**
  Prepared as described in Method A.
- Sodium hydroxide, ca. 0.1 N
  From Method A, Procedure, Step 4; exact N calculated.
- Standard buffers, pH 4.0 and 7.0

Hazards, Precautions, and Waste Disposal

Adhere to normal laboratory safety procedures. Wear safety glasses at all times. Waste likely may be put down the drain using a water rinse, but follow good laboratory practices outlined by environmental health and safety protocols at your institution.

Supplies

- Apple juice, 60 ml
- 3 Beakers, 250 ml
- 2 Burets, 25 or 50 ml
- 4 Erlenmeyer flasks, 250 ml
- Funnel, small, to fit top of 25 or 50 ml buret
- Graduated cylinder, 50 ml
- Soda, clear, 80 ml
- 2 Volumetric pipettes, 10 ml

Equipment

- Hot plate
- pH meter

Procedure

I. Soda

Do at least duplicate determinations for unboiled soda and for boiled soda sample; open soda well before use to allow escape of carbon dioxide, so sample can be pipetted.

1. Unboiled soda: Pipette 20 ml of soda into a 250-ml Erlenmeyer flask. Add ca. 50 ml CO_2-free dd water. Add 3 drops of a 1% phenolphthalein solution and titrate with standardized NaOH (ca. 0.1 N) to a faint pink color (NaOH in buret, from Method A). Record the beginning and ending volumes on the buret to determine the total volume of NaOH solution used in each titration. Observe the endpoint. Note whether the color fades.

2. Boiled soda: Pipette 20 ml of soda into a 250-ml Erlenmeyer flask. Bring the sample to boiling on a hot plate, swirling the flask often. Boil the sample only 30–60 seconds. Cool to room temperature. Add ca. 50 ml CO_2-free dd water. Add 3 drops of the phenolphthalein solution and titrate as described above. Record the beginning and ending volumes on the buret to determine the total volume of NaOH solution used in each titration. Observe the endpoint. Note whether the color fades.

II. Apple Juice

1. Standardize the pH meter with pH 7.0 and 4.0 buffers, using instructions for the pH meter you have available.
2. Prepare as described below three apple juice samples that will be compared:
   A—Apple juice (Set aside, to recall original color)
   B—Apple juice; titrate with std. NaOH
   C—Apple juice; add phenolphthalein; titrate with std. NaOH; follow pH during titration

Procedure: Into each of three (A, B, C) 250-ml beakers, pipette 20 ml of apple juice. Add ca. 50 ml CO_2-free water to each. To beaker C add 3 drops of a 1% phenolphthalein solution. Using two burets filled with the standardized NaOH solution (ca. 0.1 N), titrate Samples B and C simultaneously. Follow the pH during titration of Sample C containing phenolphthalein. (*Note*: If only one buret is available, titrate Samples B and C sequentially, i.e., add 1 ml to B, then 1 ml to C.) Record the initial pH and the pH at ca. 1.0 ml intervals until a pH of 9.0 is reached. Also observe any color changes that occur during the titration to determine when the phenolphthalein endpoint is reached. Sample A is intended to help you remember the original color of the apple juice. Sample B (without phenolphthalein) does not need to be followed with the pH meter, but is to be titrated along with the other beaker to aid in observing color changes.

Data and Calculations

I. Soda

Using the volume of NaOH used, calculate the titratable acidity (TA) of each soda sample as percentage citric acid, then calculate the mean TA of each type of sample (MW citric acid = 192.14; equivalent weight = 64.04) (Note: For equation, see Chapter 13 of Nielsen, *Food Analysis*.)

Trial	Buret Start	Buret End	Vol. NaOH Titrant	Color Fades?	TA
Unboiled soda					
1					
2					
					$\overline{X}$ =
Boiled soda					
1					
2					
					$\overline{X}$ =

Sample calculation:

% Acid =

$$\frac{(\text{ml base titrant}) \times (N \text{ of base in mol/liter}) \times (\text{Eq. wt. of acid})}{(\text{sample volume in ml}) \times 10}$$

N NaOH = 0.1019 N
ml base = 7 ml
Eq. wt. citric acid = 64.04
Vol. sample = 20 ml

% Acid = $(7 \times 0.1019 \times 64.04)/(20 \times 10)$

= 0.276% citric acid

II. Apple Juice

Sample B

Color change during titration:
Color at end of titration:

Sample C (titrate to >pH 9.0)

ml NaOH	1	2	3	4	5	6	7	8	9	10
pH										

ml NaOH	11	12	13	14	15	16	17	18	19	20
pH										

Plot pH versus ml of 0.1 N NaOH (but use the normality of your own NaOH solution) (pH on the y-axis) for the sample that contained phenolphthalein (Beaker C). Interpolate to find the volume of titrant at pH 8.2 (the phenolphthalein endpoint).

Calculate the titratable acidity of the apple juice as percentage malic acid (MW malic acid = 134.09; equivalent weight = 67.04).

Questions

1. Soda samples. (a) Did any color changes occur in either the boiled or the unboiled sample within several minutes of the phenolphthalein endpoint being reached? (b) How did boiling the sample affect the determination of titratable acidity? (c) Explain the differences in color changes and titratable acidity between the two samples.
2. What caused the color changes in the apple juice titrated without any phenolphthalein present? (Hint: Consider the pigments in apples.) How would you recommend determining the endpoint in the titration of tomato juice?
3. You are determining the titratable acidity of a large number of samples. You ran out of freshly boiled dd H_2O with an ascarite trap on the water container, so you switch to using tap distilled H_2O. Would this likely affect your results? Explain.
4. The electrode of your pH meter has a slow response time and seems to need cleaning, as it is heavily used for a variety of solutions high in proteins, lipids, and minerals. You would ideally check the electrode instructions for specific recommendations on cleaning, but the instructions were thrown away. (As the new lab supervisor, you have since started a policy of filing all instrument/equipment instructions.) What solutions would you use to try to clean the electrode?

REFERENCES

Sadler, G.D., and Murphy, P.A. 2003. pH and Titratable acidity. Ch. 13, in *Food Analysis*, 3rd ed. S.S. Nielsen (Ed.), Kluwer Academic, New York.
AOAC International. 2000. Method 942.15. *Official Methods of Analysis*, 17th ed. AOAC International, Gaithersburg, MD.

Fat Characterization

Laboratory Developed in part by

Dr Michael Quia,
Oregon State University, Department of Food Science, Corvallis, Oregon
and

Dr Oscar Pike,
Brigham Young University, Department of Nutrition, Dietetics, and Food Science, Provo, Utah

INTRODUCTION

Background

Lipids in food are subjected to many chemical reactions during processing and storage. While some of these reactions are desirable, others are undesirable so efforts are made to minimize the reactions and their effects. The laboratory deals with characterization of fats and oils with respect to composition, structure, and reactivity.

Reading Assignment

Pike, O. A. 2003. Fat characterization, Ch. 14 in *Food Analysis*, 3rd ed. S.S. Nielsen (Ed.), Kluwer Academic, New York.

Overall Objective

The overall objective of this laboratory is to determine aspects of the composition, structure, and reactivity of fats and oils by various methods.

METHOD A: SAPONIFICATION VALUE

Objective

Determine the saponification number of fats and oils.

Principle of Method

Saponification is the process of treating a neutral fat with alkali, breaking it down to glycerol and fatty acids. The saponification value (or number) is defined as the amount of alkali needed to saponify a given quantity of fat or oil, expressed as mg potassium hydroxide to saponify 1 g sample. Excess alcoholic potassium hydroxide is added to the sample, the solution is heated to saponify the fat, the unreacted potassium hydroxide is back-titrated with standardized hydrochloric acid using a phenolphthalein indicator, and the calculated amount of reacted potassium hydroxide is used to determine the saponification value.

Chemicals

	CAS No.	Hazards
Ethanol	64-17-5	Highly flammable
Hydrochloric acid (HCl)	7647-01-0	Corrosive
Phenolphthalein	77-09-8	Irritant
Potassium hydroxide (KOH)	1310-58-3	Corrosive

Reagents

(**It is recommended that these solutions be prepared by the laboratory assistant before class.)

- Alcoholic potassium hydroxide, ca. 0.7 N**
  Dissolve 40 g KOH, low in carbonate, in 1 L of distilled ethanol, keeping temperature below 15.5°C while alkali is dissolved. The solution should be clear.
- Hydrochloric acid, ca. 0.5 N, accurately standardized**
  Prepare ca. 0.5 N HCl. Determine exact normality using solution of standard base.
- Phenolphthalein indicator solution**
  1%, in 95% ethanol

Hazards, Precautions, and Waste Disposal

Use hydrochloric acid in a fume hood. Otherwise, adhere to normal laboratory safety procedures. Wear safety glasses at all times. Wastes likely may be put down the drain using a water rinse, but follow good laboratory practices outlined by environmental health and safety protocols at your institution.

Supplies

(Used by students)

- Air (reflux) condenser (650 mm long, minimum)
- Beaker, 250 ml (to melt fat)
- Buchner funnel (to fit side-arm flask)
- Boiling beads
- 2 Burets, 50 ml
- Fat and/or oil samples
- Filter paper (to fit Buchner funnel; to filter oil and melted fat)
- 4 Flasks, 250–300 ml, to fit condenser
- Mechanical pipettor, 1000 µl, with plastic tips (or 1-ml volumetric pipette)
- Side-arm flask

Equipment

- Analytical balance
- Hot plate or water bath (with variable heat control)

Procedure

(Instructions are given for analysis in duplicate.)

1. Melt any solid samples. Filter melted fat sample and oil sample through filter paper to remove impurities.
2. Weigh accurately ca. 5 g melted fat or oil into each of two 250–300 flasks that will connect to

a condenser. Record weight of sample. Prepare sample in duplicate.

3. Add accurately (from a buret) 50 ml of alcoholic KOH into the flask.

4. Prepare duplicate blank samples with just 50 ml of alcoholic KOH in a 250–300 ml flask.

5. Add several boiling beads to flasks with fat or oil sample.

6. Connect the flasks with sample to a condenser. Boil gently but steadily on a hot plate (or water bath) until sample is clear and homogenous, which indicates complete saponification (requires ca. 30–60 min). (Note: The fumes should condense as low as possible in the condenser, otherwise a fire hazard is created.)

7. Allow samples to cool somewhat. Wash down the inside of the condenser with a little deionized distilled (dd) water. Disconnect flask from condenser. Allow sample to cool to room temperature.

8. Add 1 ml phenolphthalein to samples and titrate with 0.5 N HCl (from a buret) until pink color has just disappeared. Record the volume of titrant used.

9. Repeat Steps 5–8 with sample blanks. Reflux the blanks for the same time period as used for the sample.

Data and Calculations

Sample	Weight (g)	Titrant Volume (ml)	Saponification Value
1			
2			
		$\overline{X} =$	

Oil/fat sample type tested:

Blank Titration (ml)
    Sample 1 =
    Sample 2 =
        $\overline{X} =$

Calculate the saponification number (or value) of each sample as follows:

$$\text{Saponification value} = \frac{(B - S) \times N \times 56.1}{W}$$

where:
    Saponification value = mg KOH per g of sample
    B = volume of titrant (ml) for blank
    S = volume of titrant (ml) for sample
    N = normality of HCl (mmol/ml)
    56.1 = molecular weight (MW) of KOH (mg/mmol)
    W = sample mass (g)

Questions

1. What is meant by unsaponifiable matter in lipid samples? Give an example of such a type of compound.
2. What does a high versus a low saponification value tell you about the name of a sample?

METHOD B: IODINE VALUE

Objective

Determine the iodine value of fats and oils.

Principle of Method

The iodine value (or number) is a measure of the degree of unsaturation, defined as the grams of iodine absorbed per 100-g sample. In the assay, a measured quantity of fat or oil dissolved in solvent is reacted with a measured excess amount of iodine or some other halogen, which reacts with the carbon–carbon double bonds. After a solution of potassium iodide is added to reduce excess ICl to free iodine, the liberated iodine is titrated with a standardized solution of sodium thiosulfate using a starch indicator. The calculated amount of iodine reacted with the double bonds is used to calculate the iodine value.

Chemicals

	CAS No.	Hazards
Acetic acid (glacial)	64-19-7	Corrosive
Carbon tetrachloride (CCl_4)	56-23-5	Toxic, danger for the environment
Chloroform	67-66-3	Harmful
Hydrochloric acid (HCl)	7647-01-0	Corrosive
Iodine	7553-56-2	Harmful, dangerous for the environment
Potassium chromate ($K_2Cr_2O_7$)	7789-00-6	Toxic, dangerous for the environment
Potassium iodide (KI)	7681-11-0	
Sodium thiosulfate	7772-98-7	
Soluble starch	9005-25-8	

Reagents

(**It is recommended that these solutions be prepared by the laboratory assistant before class.)

- Potassium iodide solution, 15% Dissolve 150 g KI in dd water and dilute to 1 liter.
- Sodium thiosulfate, 0.1 N standardized solution (AOAC Method 942.27)** Dissolve ca. 25 g sodium thiosulfate in 1 L dd water. Boil gently for 5 min. Transfer while hot to a storage bottle (make sure bottle has been well

cleaned, and is heat resistant). Store solution in a dark, cool place. Use the following procedure to standardize the sodium thiosulfate solution: Accurately weigh 0.20–0.23 g potassium chromate ($K_2Cr_2O_7$) (previously dried for 2 hr at 100°C) into a glass-stoppered flask. Dissolve 2 g potassium iodide (KI) in 80 ml chlorine-free water. Add this water to the potassium chromate. To this solution, add, with swirling, 20 ml ca. 1 M HCl, and immediately place in the dark for 10 min. Titrate a known volume of this solution with the sodium thiosulfate solution, adding starch solution after most of the iodine has been consumed.

- Starch indicator solution, 1% (prepare fresh daily) Mix ca. 1 g soluble starch with enough cold dd water to make a thin paste. Add 100 ml boiling dd water. Boil ca. 1 min while stirring.

- Wijs iodine solution**
  Dissolve 10 g ICl_3 in 300 ml CCl_4 and 700 ml glacial acetic acid. Standardize this solution against 0.1 N sodium thiosulfate (25 ml of Wijs solution should consume 3.4–3.7 mEq of thiosulfate). Then, add enough iodine to the solution such that 25 ml of the solution will require at least 1.5 times the milliequivalency of the original titration. Place the solution in an amber bottle. Store in the dark at less than 30°C.

Hazards, Precautions, and Waste Disposal

Carbon tetrachloride and potassium chromate are toxic and must be handled with caution. Use acetic acid and hydrochloric acid in a fume hood. Otherwise, adhere to normal laboratory safety procedures. Wear safety glasses at all times. Carbon tetrachloride, chloroform, iodine, and potassium chromate must be handled as hazardous wastes. Other wastes likely may be put down the drain using a water rinse, but follow good laboratory practices outlined by environmental health and safety protocols at your institution.

Supplies

(Used by students)

- 2 Beakers, 250 ml (one to melt fat; one to boil water)
- Buchner funnel (to fit side-arm flask)
- Buret, 10 or 25 ml
- Fat and/or oil samples
- Filter paper (to fit Buchner funnel; to filter melted fat and oil)
- 4 Flasks, 500 ml, glass-stoppered
- Graduated cylinder, 25 ml
- Graduated cylinder, 100 ml

- Mechanical pipettor, 1000 μl, with plastic tips (or 1-ml volumetric pipette)
- Side-arm flask
- Volumetric pipette, 10 ml
- Volumetric pipette, 20 ml

Equipment

- Analytical balance
- Hot plate

Procedure

(Instructions are given for analysis in duplicate.)

1. Melt any samples that are solid at room temperature by heating to a maximum of 15°C above the melting point. Filter melted fat sample and oil sample through filter paper to remove impurities.
2. Weigh accurately 0.1 to 0.5 g sample (amount used depends on expected iodine number) into each of two dry 500-ml glass-stoppered flasks. Add 10 ml chloroform to dissolve the fat or oil.
3. Prepare two blanks by adding only 10 ml chloroform to 500-ml glass-stoppered flasks.
4. Pipette 25 ml Wijs iodine solution into the flasks. (The amount of iodine must be 50–60% in excess of that absorbed by the fat.)
5. Let flasks stand for 30 min in the dark with occasional shaking.
6. After incubation in the dark, add 20 ml potassium iodide solution to each flask. Shake thoroughly. Add 100 ml freshly boiled and cooled water, washing down any free iodine on the stopper.
7. Titrate the iodine in the flasks with standard sodium thiosulfate, adding it gradually with constant and vigorous shaking until the yellow color almost disappears. Then add 1–2 ml of starch indicator and continue the titration until the blue color entirely disappears. Toward the end of the titration, stopper the flask and shake violently so that any iodine remaining in the chloroform can be taken up by the potassium iodide solution. Record the volume of titrant used.

Data and Calculations

Sample	Weight (g)	Titrant Volume (ml)	Iodine Value
1			
2			
			$\overline{X} =$

Oil/fat sample type tested:

Blank Titration (ml)
      Sample 1 =
      Sample 2 =
      $\overline{X}$ =

Calculate the iodine value of each sample as follows:

$$\text{Iodine value} = \frac{(B - S) \times N \times 126.9}{W} \times 100$$

where:

$$
\begin{aligned}
\text{Iodine value} &= \text{g iodine absorbed per } 100\,\text{g of} \\
&\quad \text{sample} \\
B &= \text{volume of titrant (ml) for blank} \\
S &= \text{volume of titrant (ml) for sample} \\
N &= \text{normality of } Na_2S_2O_3 \\
&\quad (\text{mol}/1000\,\text{ml}) \\
126.9 &= \text{MW of iodine (g/mol)} \\
W &= \text{sample mass (g)}
\end{aligned}
$$

Questions

1. In the iodine value determination, why is the blank volume higher than that of the sample?
2. What does a high versus a low iodine value tell you about the nature of the sample?

METHOD C: FREE FATTY ACID VALUE

Objective

Determine the free fatty acid (FFA) value of fats and oils.

Principle of Method

Free fatty acid value, or acid value, reflects the amount of fatty acids hydrolyzed from triacylglycerols. Free fatty acid is the percentage by weight of a specific fatty acid. Acid value is defined as the milligrams of potassium hydroxide needed to neutralize the free acids present in 1 g of fat or oil. A liquid fat sample combined with neutralized 95% ethanol is titrated with standardized sodium hydroxide to a phenolphthalein endpoint. The volume and normality of the sodium hydroxide are used, along with the weight of the sample, to calculate the free fatty acid value.

Chemicals

	CAS No.	Hazards
Ethanol	64-17-5	Highly flammable
Phenolphthalein	77-09-8	Irritant
Sodium hydroxide (NaOH)	1310-73-2	Corrosive

Reagents

(**It is recommended that these solutions be prepared by the laboratory assistant before class.)

- Ethanol, neutralized
  Neutralize 95% ethanol to a permanent pink color with alkali and phenolphthalein.
- Phenolphthalein indicator
  In a 100-ml volumetric flask, dissolve 1 g phenolphthalein in 50 ml 95% ethanol. Dilute to volume with dd water.
- Sodium hydroxide, 0.1 N, standardized**
  Use commercial product, or prepare as described in laboratory experiment, "Standard Solutions and Titratable Acidity," Chapter 11 (above), Method A.

Hazards, Precautions, and Waste Disposal

Adhere to normal laboratory safety procedures. Wear safety glasses at all times. Wastes likely may be put down the drain using a water rinse, but follow good laboratory practices outlined by environmental health and safety protocols at your institution.

Supplies

(Used by students)

- Beaker, 250 ml (to melt fat)
- Buchner funnel (to fit side-arm flask)
- Buret, 10 ml
- 4 Erlenmeyer flasks, 250 ml
- Fat and/or oil samples
- Filter paper (to fit Buchner funnel; to filter melted fat and oil)
- Graduated cylinder, 100 ml
- Mechanical pipettor, 1000 µl, with plastic tips (or 1-ml volumetric pipette)
- Side-arm flask

Equipment

- Analytical balance
- Hot plate

Procedure

(Instructions are given for analysis in triplicate.)

1. Melt any samples that are solid at room temperature by heating to a maximum of 15°C above the melting point. Filter melted fat sample and oil sample through filter paper to remove impurities.
2. As a preliminary test, accurately weigh ca. 5 g melted fat or oil into a 250-ml Erlenmeyer flask.

3. Add ca. 100 ml neutralized ethanol and 2 ml phenolphthalein indicator.
4. Shake to dissolve the mixture completely. Titrate with standard base (ca. $0.1\,N$ NaOH), shaking vigorously until the endpoint is reached. This is indicated by a slight pink color that persists for 30 s. Record the volume of titrant used. Use the information below to determine if the sample weight you have used is correct for the range of acid values under which your sample falls. This will determine the sample weight to be used for Step 5.

   The *Official Methods and Recommended Practices of the AOCS* (AOCS 1998) recommends the following sample weights for ranges of expected acid values:

FFA Range (%)	Sample (g)	Alcohol (ml)	Strength of Alkali
0.00 to 0.2	56.4 ± 0.2	50	$0.1\,N$
0.2 to 1.0	28.2 ± 0.2	50	$0.1\,N$
1.0 to 30.0	7.05 ± 0.05	75	$0.25\,N$

5. Repeat Steps 1–3 more carefully in triplicate, recording each weight of the sample and the volume of titrant.

Data and Calculations

Sample	Weight (g)	Titrant Volume (ml)	FFA Value
1			
2			
3			
			$\overline{X} =$
			SD $=$

Oil/fat sample type tested:

Calculate the FFA value of each sample as follows:

$$\% \text{ FFA (as oleic)} = \frac{V \times N \times 282}{W} \times 100$$

where:
   $\%$ FFA $=$ Percent free fatty acid (g/100 g), expressed as oleic acid
   $V =$ Volume of NaOH titrant (ml)
   $N =$ Normality of NaOH titrant (mol/1000 ml)
   $282 =$ MW of oleic acid (g/mol)
   $W =$ sample mass (g)

Questions

1. What is a high FFA value indicative of relative to product history?
2. Why is the FFA content of frying oil important?

3. In a crude fat extract, FFA are naturally present, but they are removed during processing to enhance the stability of the fat. State and describe the processing step that removes the FFA naturally present.

METHOD D: PEROXIDE VALUE

Objective

Determine the peroxide value of fats and oils, as an indicator of oxidative rancidity.

Principle of Method

Peroxide value is defined as the milliequivalents of peroxide per kilogram of fat, as determined in a titration procedure to measure the amount of peroxide or hydroperoxide groups. To a known amount of fat or oil, excess potassium iodide is added, which reacts with the peroxides in the sample. The iodine liberated is titrated with standardized sodium thiosulfate using a starch indicator. The calculated amount of potassium iodide required to react with the peroxide present is used to determine the peroxide value.

Chemicals

	CAS No.	Hazards
Acetic acid (glacial)	64-19-7	Corrosive
Chloroform	67-66-3	Harmful
Hydrochloric acid (HCl)	7647-01-0	Corrosive
Potassium chromate ($K_2Cr_2O_7$)	7789-00-6	Toxic, dangerous for environment
Potassium iodide (KI)	7681-11-0	
Sodium thiosulfate	7772-98-7	
Soluble starch	9005-25-8	

Reagents

(**It is recommended that these solutions be prepared by the laboratory assistant before class.)

- Acetic acid-chloroform solution
  Mix 3 volumes of concentrated acetic acid with 2 volumes of chloroform.
- Potassium iodide solution, saturated**
  Dissolve excess KI in freshly boiled dd water. Excess solid must remain. Store in the dark. Test before use by adding 0.5 ml acetic acid-chloroform solution, then add 2 drops 1% starch indicator solution. If solution turns blue, requiring >1 drop $0.1\,N$ thiosulfate solution to discharge color, prepare a fresh potassium iodide solution.

- Sodium thiosulfate, 0.2 N, standard solution (AOAC Method 942.27)**
  Dissolve ca. 50 g sodium thiosulfate in 1 liter dd water. Boil gently for 5 min. Transfer while hot to a storage bottle (make sure bottle has been well cleaned, and is heat resistant). Store solution in a dark, cool place. Use the following procedure to standardize the sodium thiosulfate solution: Accurately weigh 0.20–0.23 g potassium chromate ($K_2Cr_2O_7$) (previously dried for 2 hr at 100°C) into a glass-stoppered flask. Dissolve 2 g potassium iodide (KI) in 80 ml chlorine-free water. Add this water to the potassium chromate. To this solution, add, with swirling, 20 ml ca. 1 M HCl, and immediately place in the dark for 10 min. Titrate a known volume of this solution with the sodium thiosulfate solution, adding starch solution after most of the iodine has been consumed.
- Starch indicator solution, 1% (prepare fresh daily)
  Mix ca. 1 g soluble starch with enough cold dd water to make a thin paste. Add 100 ml boiling dd water. Boil ca. 1 min while stirring.

Hazards, Precautions, and Waste Disposal

Potassium chromate is toxic and must be handled with caution. Use hydrochloric acid in a fume hood. Otherwise, adhere to normal laboratory safety procedures. Wear gloves and safety glasses at all times. Chloroform and potassium chromate must be handled as hazardous wastes. Other wastes likely may be put down the drain using a water rinse, but follow good laboratory practices outlined by environmental health and safety protocols at your institution.

Supplies

(Used by students)

- Beaker, 250 ml (to melt fat)
- Buchner funnel (to fit side-arm flask)
- Buret, 25 ml or 50 ml
- 4 Erlenmeyer flasks, 250 ml, glass stoppered
- Fat and/or oil samples
- Filter paper (to fit Buchner funnel; to filter melted fat and oil)
- 2 Graduated cylinders, 50 ml
- Mechanical pipettor, 1000 μl, with plastic tips (or 1 ml volumetric pipette)
- Side-arm flask

Equipment

- Analytical balance
- Hot plate

Procedure

(Instructions are given for analysis in duplicate.)

1. Melt any samples that are solid at room temperature by heating to a maximum of 15°C above the melting point. Filter melted fat sample and oil sample through filter paper to remove impurities.
2. Accurately weigh ca. 5 g fat or oil (to the nearest 0.001 g) into each of two 250-ml glass-stoppered Erlenmeyer flasks.
3. Add 30 ml acetic acid-chloroform solution and swirl to dissolve.
4. Add 0.5 ml saturated KI solution. Let stand with occasional shaking for 1 min. Add 30 ml dd water.
5. Slowly titrate samples with 0.1 N sodium thiosulfate solution, with vigorous shaking until yellow color is almost gone.
6. Add ca. 0.5 ml 1% starch solution, and continue titration, shaking vigorously to release all iodine from chloroform layer, until blue color just disappears. Record the volume of titrant used. (If <0.5 ml of the sodium thiosulfate solution is used, repeat determination.)
7. Prepare (omitting only the oil) and titrate a blank sample. Record the volume of titrant used.

Data and Calculations

Sample	Weight (g)	Titrant Volume (ml)	Peroxide Value
1			
2			
			$\overline{X}=$

Oil/fat sample type tested:

Blank Titration (ml)
  Sample 1 =
  Sample 2 =
  $\overline{X}=$

Calculate the peroxide value of each sample as follows:

$$\text{Peroxide value} = \frac{(S - B) \times N}{W} \times 1000$$

where:
  Peroxide value = mEq peroxide per kg of sample
  S = volume of titrant (ml) for sample
  B = volume of titrant (ml) for blank
  N = normality of $Na_2S_2O_3$ solution (mEq/ml)
  1000 = conversion of units (g/kg)
  W = sample mass (g)

Questions

1. What are some cautions in using peroxide value to estimate the amount of autoxidation in foods?
2. The peroxide value method was developed for fat or oil samples. What must be done to a food sample before measuring its peroxide value using this method?

METHOD E: THIN-LAYER CHROMATOGRAPHY SEPARATION OF SIMPLE LIPIDS

Objective

Separate and identify the lipids in some common foods using thin-layer chromatography (TLC).

Principle of Method

Like all types of chromatography, TLC is a separation technique that allows for the distribution of compounds between a mobile phase and a stationary phase. Most classes of lipids can be separated from each other by adsorption chromatography on thin layers. In TLC, a thin layer of stationary phase is bound to an inert support (i.e., glass plate, plastic or aluminum sheet). The sample and standards are applied as spots near one end of the plate. For ascending chromatography, the plate is placed in a developing chamber, with the end of the plate nearest the spots being placed in the mobile phase at the bottom of the chamber. The mobile phase migrates up the plate by capillary action, carrying and separating the sample components. The separated bands can be visualized or detected, and compared to the separation of standard compounds.

Chemicals

	CAS No.	Hazards
Acetic acid	64-19-7	Corrosive
Diethyl ether	60-29-7	Harmful, extremely flammable
Hexane	110-54-3	Harmful, highly flammable, dangerous for the environment
Sulfuric acid	7664-93-9	Corrosive

Reagents

- Chloroform : methanol, 2 : 1, v/v
- Mobile phase
  Hexane : diethyl ether : acetic acid, 78 : 20 : 2
- Standards
  Triacylglycerol, fatty acid, cholestyrl ester, and cholesterol
- Sulfuric acid solution
  Concentrated H_2SO_4, in 50% aqueous solution

Hazards, Precautions, and Waste Disposal

Use acetic acid and sulfuric acid in a fume hood. Diethyl ether is extremely flammable, is hygroscopic, and may form explosive peroxides. Otherwise, adhere to normal laboratory safety procedures. Wear safety glasses at all times. Diethyl ether and hexane must be handled as hazardous wastes. Other wastes likely may be put down the drain using a water rinse, but follow good laboratory practices outlined by environmental health and safety protocols at your institution.

Supplies

- Capillary tubes (or syringes) (to apply samples to plates)
- Developing tank, with lid
- Filter paper, Whatman No. 1 (to line developing tank)
- Oil/fat food samples (e.g., hamburger, safflower oil) (prepare at a concentration of 20 μg/ml in 2 : 1 v/v chloroform-methanol solution)
- Pencil
- Thin-layer chromatography plates: Silica Gel 60, 0.25 mm thick coating on glass backing, 20 × 20 cm (EM Science)

Equipment

- Air blower (e.g., blow hair dryer)
- Oven

Procedure

I. Preparation of Silica Gel Plates

1. Place plates in oven at 110°C for 15 min, then cool to ambient temperature (5 min).
2. With a pencil, draw a line to mark the origin, 2.5 cm from the bottom of the plate.
3. Make marks with a pencil to divide the plate into 10 "lanes" of equal width.
4. Use capillary tubes or syringes to apply approximately 10 μl of each standard and sample to a separate lane (use the middle eight lanes). The application should be done as a streak across the center of the lane origin. This is best accomplished with four spots of 2.5 μl each.
5. Below the origin line, write the identity of the sample/standard in each lane.
6. Allow spots to dry. You may accelerate drying by using a low-temperature air blower.
7. Write your name in the top right corner of the plate.

II. Development of Plates

1. Line the developing tank with Whatman no. 1 or similar filter paper.
2. Pour the mobile phase gently over the filter paper until the depth of solvent in the tank is approximately 0.5 cm. About 200 ml is required.
3. Place the lid on the tank and allow 15 min for the atmosphere in the tank to become saturated with solvent vapor.
4. Place the spotted TLC plate in the developing tank and allow it to develop until the solvent front reaches a point about 2 cm from the top of the plate.
5. Remove the plate from the tank and *immediately* mark the position of the solvent front. Evaporate the solvent in the fume hood.

III. Visualization of Lipids

1. In a well-ventilated fume hood spray lightly with 50% aqueous H_2SO_4. Allow to dry.
2. Heat plate for 5–10 min at 100–120°C. Remove from oven, cool, and inspect. Handle the plate with caution as the surface still contains sulfuric acid.
3. Mark all visible spots at their center, and note the color of the spots.

Data and Calculations

For the spots of each of the standards and the samples, report the distance from the origin for the spot. Also for each spot, calculate the R_f value, as the distance from the origin to the spot divided by the distance from the origin to the solvent front. Using the R_f value of the standards, identify as many of the spots (bands) in the samples as possible.

Standard	Distance from Origin	R_f Value
Triacylglycerol		
Fatty acid		
Cholestyrl ester		
Cholesterol		

Sample Spot Number	Distance from Origin	R_f Value	Identity

Oil/fat sample(s) type tested:

Questions

1. Explain the chemical structure of an ester of cholesterol.
2. Besides the four fat constituents used as standards, what other fat constituents might be found using a TLC method such as this?

ACKNOWLEDGMENTS

This laboratory exercise was developed with input from Dr Arun Kilara with Arun Kilara Worldwide, Northbrook, IL.

REFERENCES

AOCS. 1998. *Official Methods and Recommended Practices of the AOCS*, 5th ed. American Oil Chemists' Society, Champaign, IL.

Pike, O.A. 2003. Fat characterization, Ch. 14 in *Food Analysis*, 3rd ed. S.S. Nielsen (Ed.), Kluwer Academic, New York.

Fish Muscle Proteins: Extraction, Quantitation, and Electrophoresis

Laboratory Developed in part by

Dr Denise M. Smith,
University of Idaho, Department of Food Science and Toxicology,
Moscow, Idaho

INTRODUCTION

Background

Electrophoresis can be used to separate and visualize proteins. In sodium dodecyl sulfate-polyacrylamide gel electrophoresis (SDS-PAGE), proteins are separated based on size. When proteins samples are applied to such gels, it is usually necessary to know the protein content of the sample. This makes it possible to apply a volume of sample to the gel so samples have a comparable amount of total protein. While it is possible to use an official method of protein analysis (e.g., Kjeldahl, N combustion) for such an application, it often is convenient to use a rapid protein analysis that requires only a small amount of sample. In this experiment, the bicinchoninic acid (BCA) method will be used for this purpose.

In this experiment, sarcoplasmin muscle proteins are extracted with a salt solution, the protein content of the extract is measured by the BCA assay, and the proteins in the fish extracts are separated and visualized by SDS-PAGE. This visualization of the proteins makes it possible to distinguish different types of fish since most fish have a characteristic protein pattern. One can detect the substitution of inexpensive fish for an expensive fish.

Reading Assignment

Chang, S.K.C. 2003. Protein analysis. Ch. 9, in *Food Analysis*, 3rd ed. S.S. Nielsen (Ed.), Kluwer Academic, New York.

Smith, D.M. 2003. Protein separation and characterization. Ch. 15, in *Food Analysis*, 3rd ed. S.S. Nielsen (Ed.), Kluwer Academic, New York.

Objective

Extract proteins from the muscles of freshwater and saltwater fish, measure the protein content of the extracts, then separate and compare the proteins by electrophoresis.

Principle of Method

Sarcoplasmic proteins can be extracted from fish muscle with salt. Protein content of the extract can be determined by the BCA method, in which protein present reduces cupric ions to cuprous ions under alkaline conditions. The cuprous ions react with BCA reagent to give a purple color that can be quantitated spectrophotometrically and related to protein content. Proteins present in the extract can be separated by SDS-PAGE, which gives a size separation. Proteins bind SDS to become negatively charged, so they move through the gel matrix toward the anode (pole with positive charge)

at a rate based on size alone, since all molecules are highly negatively charged. The molecular weight of a given protein can be estimated by comparing its electrophoretic mobility with proteins of known molecular weight. A linear relationship is obtained if the logarithm of the molecular weights of standard proteins are plotted against their respective electrophoretic mobilities (R_f).

Notes

This experiment may be done over two laboratory sessions, with protein extracted and quantitated in the first session. The protein samples can be frozen after preparation for electrophoresis, for running on gels in the second laboratory session. Alternatively, in a single laboratory session, one group of students could do the protein extraction and quantitation, while a second group of students prepares the electrophoresis gels. Also, different groups of students could do extraction and quantitation of the saltwater fish and the freshwater fish. Multiple groups could run their samples on a single electrophoresis gel. The gels for electrophoresis can be purchased commercially (e.g., BioRad, PROTEAN II Ready Gel Precast Gels, 15% resolving gel, 4% stacking gel), rather than made as described below.

Some fish species work better than others for preparing the extracts, and showing a difference between freshwater and saltwater fish. Catfish (freshwater) and tilapia (saltwater) work well as extracts and show some differences. Trout gives very thick extracts. Freshwater and saltwater salmon show few differences.

Chemicals

	CAS No.	Hazards
Sample Extraction		
Sodium chloride (NaCl)	7647-14-5	Irritant
Sodium phosphate, monobasic (NaH$_2$PO$_4$ · H$_2$O)	7558-80-7	Irritant
Protein Determination (*BCA method*)		
Bicinchoninic acid		
Bovine serum albumin (BSA)	9048-46-8	
Copper sulfate (CuSO$_4$)	7758-98-7	Irritant
Sodium bicarbonate (NaHCO$_3$)	144-55-8	
Sodium carbonate (Na$_2$CO$_3$)	497-19-8	Irritant
Sodium hydroxide (NaOH)	1310-73-2	Corrosive
Sodium tartrate	868-18-8	
Electrophoresis		
Acetic acid (CH$_3$COOH)	64-19-7	Corrosive
Acrylamide	79-06-1	Toxi
Ammonium persulfate (APS)	7727-54-0	Harmful, oxidizing

(Continued)

	CAS no.	Hazards
Bis-acrylamide	110-26-9	Harmful
Butanol	71-36-3	Harmful
Coomassie Blue R-250	6104-59-2	
Ethylenediaminetetraacetic acid, disodium salt (Na$_2$EDTA· 2H$_2$O)	60-00-4	Irritant
Glycerol (C$_3$H$_8$O$_3$)	56-81-5	
Glycine	56-40-6	
Hydrochloric acid (HCl)	7647-01-0	Corrosive
β-Mercaptoethanol	60-24-2	Toxic
Methanol (CH$_3$OH)	67-56-1	Highly flammable
Protein molecular weight standards (e.g., BioRad 161-0318, Prestained SDS-PAGE standards broad range, 209–7 kD)		
Sodium dodecyl sulfate (SDS, Dodecyl sulfate, sodium salt)	151-21-3	Harmful
N, N, N', N'-Tetramethylethylene-diamine (TEMED)	110-18-9	Highly flammable, corrosive
Tris base	77-86-1	

REAGENTS

(**It is recommended that these solutions be prepared by laboratory assistant before class.)

Sample Extraction

- Extraction buffer, 300 ml/fish species**
  Buffer of 0.15 M sodium chloride, 0.05 M sodium phosphate, pH 7.0. (Students asked to show calculations for this buffer later in Questions.)

Protein Determination (BCA method)

[Purchase Reagents A and B from Pierce (Rockford, IL).]

- Bovine serum albumin (BSA) standard solution, 1 mg/ml**
  Accurately weigh 1 mg BSA. Place in small test tube, and dissolve with 1 ml deionized distilled (dd) water.
- Reagent A: Contains sodium carbonate, sodium bicarbonate, BCA detection reagent, and sodium tartrate in 0.2 N sodium hydroxide.
- Reagent B: 4% copper sulfate solution

Electrophoresis

(*Note*: Many of these solutions can be purchased commercially. β-Mercaptoethanol may need to be added to a commercial sample preparation buffer.)

- Acrylamide : Bis-Acrylamide Solution**
  29.2 g acrylamide and 2.4 g methylene bis-acrylamide, with dd water to 100 ml.
- Ammonium Persulfate (APS), 7.5%, in dd water, 1 ml, prepared fresh daily**
- Bromophenol Blue, 0.05%
- Coomassie Brilliant Blue Stain Solution**
  454 ml dd water, 454 ml methanol, 92 ml acetic acid, and 1.5 g Coomassie Brilliant Blue R 250 (Sigma).
- Destain Solution**
  454 ml dd water, 454 ml methanol, 92 ml acetic acid.
- EDTA, disodium salt, 0.2 M, 50 ml**
- Glycerol, 37% (use directly)
- Sample Preparation Buffer**
  1 ml of 0.5 M Tris (pH 6.8), 0.8 ml glycerol, 1.6 ml 10% SDS, 0.4 ml β-mercaptoethanol, and 0.5 ml 0.05% (wt/vol) bromophenol blue, diluted to 8 ml with dd water.
- Sodium Dodecyl Sulfate, 10% solution in dd water, 10 ml**
- TEMED (use directly)
- Tray Buffer**
  15 g Tris base, 43.2 g glycine, 3 g SDS, in 1 L dd water.
- Tris Buffer, 1.5 M, pH 8.8, 50 ml (separating gel buffer)**
  Buffered to pH 8.8 by adding concentrated HCl dropwise over a 15–30 min period.
- Tris Buffer, 0.5 M, pH 6.8, 50 ml (stacking gel buffer)**
  Buffered to pH 6.8 by adding concentrated HCl dropwise.

Gel Preparation: Use formula that follows, and instructions in Procedure.

Formula for two 8.4 × 5.0 cm SDS-PAGE slab gels, 15% acrylamide, 0.75 mm thick:

Reagent	Separating Gel 15% gel	Stacking Gel 4.5% gel
Acrylamide : Bisacrylamide	2.4 ml	0.72 ml
10% SDS	80 μl	80 μl
1.5 M Tris, pH 8.8	2.0 ml	—
0.5 M Tris, pH 6.8	—	2.0 ml
dd Water	3.6 ml	5.3 ml
37% glycerol	0.15 ml	—
10% APS[1]	40 μl	40 μl
TEMED	10 μl	10 μl

[1] APS added to separating and stacking gels after all other reagents are combined, solution is degassed, and each gel is ready to be poured.

Hazards, Precautions, and Waste Disposal

Acrylamide may cause cancer and is very toxic in contact with skin and if swallowed. β-Mercaptoethanol is harmful if swallowed, toxic in contact with skin, and irritating to eyes. Adhere to normal laboratory safety procedures. Wear gloves and safety glasses at all times. Acrylamide and β-mercaptoethanol wastes must be disposed of as hazardous wastes. Gloves and pipette tips in contact with acrylamide and β-mercaptoethanol also should be handled as hazardous wastes. Other waste likely can be washed down the drain with a water rinse, but follow good laboratory practices outlined by environmental health and safety protocols at your institution.

SUPPLIES

(Used by students)

Sample Extraction

- Beaker, 250 ml
- Centrifuge tubes, 50 ml
- Cutting board
- Erlenmeyer flask, 125 ml
- Graduated cylinder, 50 ml
- Filter paper, Whatman No. 1
- Fish, freshwater (e.g., catfish) and saltwater species (e.g., tilapia)
- Funnel
- Knife
- Pasteur pipets and bulbs
- Test tube with cap
- Weighing boat

Protein Determination (BCA method)

- Beaker, 50 ml
- Graduated cylinder, 25 ml
- Mechanical, adjustable volume pipettor, 1000 μl, with plastic tips
- Test tubes

Electrophoresis

- Beaker, 250 ml (for boiling samples)
- 2 Erlenmeyer flasks, 2 liter (for stain and destain solutions)
- Glass boiling beads (for boiling samples)
- Graduated cylinder, 100 ml
- Graduated cylinder, 500 ml
- Hamilton syringe (to load samples on gels)
- Mechanical, adjustable volume pipettors, 1000 μl, 100 μl, and 20 μl, with plastic tips
- Pasteur pipettes, with bulbs
- Rubber stopper (to fit 25-ml side-arm flasks)
- 2 Side-arm flasks, 25 ml

- Test tubes, small size, with caps
- Tubing (to attach to vacuum system to degas gel solution)
- Weigh paper/boats

Equipment

- Analytical balance
- Aspirator system (for degassing solutions)
- Blender
- Centrifuge
- Electrophoresis unit
- pH meter
- Power supply
- Spectrophotometer
- Top loading balance
- Vortex mixer
- Water bath

PROCEDURE

(Single sample extracted.)

Sample Preparation

1. Coarsely cut up about 100 g fish muscle (representative sample) with a knife. Accurately weigh out 90 g on a top loading balance.
2. Blend 1 part fish with 3 parts extraction buffer (90 g fish and 270 ml extraction buffer) for 1.0 min in a blender. (Note: Smaller amounts of fish and buffer, but in the same 1 : 3 ratio, can be used for a small blender.)
3. Pour 30 ml of the muscle homogenate into a 50-ml centrifuge tube. Label tube with tape. Balance your tube against a classmate's sample. Use a spatula or Pasteur pipette to adjust tubes to an equal weight.
4. Centrifuge the samples at 2000 × g for 15 min at room temperature. Collect the supernatant.
5. To filter a portion of the supernatant, set a small funnel in a test tube. Place a piece of Whatman No. 1 filter paper in the funnel and moisten it with the extraction buffer. Filter the supernatant from the centrifuged sample. Collect about 10 ml of filtrate in a test tube. Cap the tube.
6. Determine protein content of filtrate using the BCA method and prepare sample for electrophoresis (see below).

BCA Protein Assay

(Instructions are given for duplicate analysis of each concentration of standard and sample.)

1. Prepare the Working Reagent for the BCA assay by combining Pierce Reagent A with Pierce

Reagent B, 50 : 1 (v/v), A : B. Use 25 ml Reagent A and 0.5 ml Reagent B to prepare 25.5 ml Working Reagent, which is enough for the BSA standard curve and testing the extract from one type of fish. (Note: This volume is adequate for assaying duplicates of five standard samples and two dilutions of each of two types of fish.)

2. Prepare the following dilutions of the supernatant (filtrate from Procedure, Sample Preparation, Step 5): dilutions of 1 : 5, 1 : 10, and 1 : 20 in *extraction buffer*. Mix well.

3. In test tubes, prepare duplicates of each reaction mixture of diluted extracts and BSA standards (using 1 mg BSA/ml solution) as indicated in the table that follows.

$$1 \frac{mg}{ml} \times 50 \mu l \times \frac{1 ml}{1000 \mu l} = 0.050 mg$$

Tube Identity	dd Water (μl)	BSA Std. (μl)	Fish Extract (μl)	Working Reagent (ml)
Blank	50	0	—	1.0
Std. 1	40	10	—	1.0
Std. 2	30	20	—	1.0
Std. 3	20	30	—	1.0
Std. 4	10	40	—	1.0
Std. 5	0	50	—	1.0
Sample 1 : 5	25	—	25	1.0
Sample 1 : 10	25	—	25	1.0
Sample 1 : 20	25	—	25	1.0

4. Mix each reaction mixture with a vortex mixer, then incubate in a water bath at 37°C for 30 min.

5. Read the absorbance of each tube at 560 nm using a spectrophotometer.

6. Use the data from the BSA samples to create a standard curve of absorbance at 562 nm versus μg protein/50 μl. Determine the equation of the line for the standard curve. Calculate the protein concentration (μg/ml) of the extract from each fish species using the equation of the line from the BSA standard curve and the absorbance value for a dilution of the fish extract that had an absorbance near the middle point on the standard curve. Remember to correct for dilution used.

7. For each type of fish extract, calculate the volumes (μl) that contain 20 and 40 μg protein. (These volumes of extract will be applied to the electrophoresis gel.)

Electrophoresis

1. Assemble the electrophoresis unit according to manufacturer's instructions, getting plates ready to pour the separating and stacking gels.

2. Use the table that follows the list of electrophoresis reagents to combine appropriate amounts of all reagents for the *separating* gel, except APS, in a side-arm flask. Degas the solution, then add APS. Proceed immediately to pour the solution between the plates to create the separating gel. Pour the gel to a height approximately 1 cm below the bottom of the sample well comb. Immediately add a layer of butanol across the top of the separating gel, adding it carefully so as not to disturb the upper surface of the separating gel. (This butanol layer will prevent a film from forming and helps obtain an even surface.) Allow the separating gel to polymerize for 30 min, then remove the butanol layer just before the stacking gel is ready to be poured.

3. Use the table that follows the list of electrophoresis reagents to combine appropriate amounts of all reagents for the *stacking* gel, except APS, in a side-arm flask. Degas the solution for 15 min (per manufacturer's instructions), then add APS. Proceed immediately to pour the solution between the plates to create the stacking gel. Immediately place the well comb between the plates and into the stacking gel. Allow the stacking gel to polymerize for 30 min before removing the well comb. Before loading the samples into the wells, wash the wells twice with dd water.

4. Mix the fish extract samples well (filtrate from Procedure, Sample Preparation, Step 5), then for each sample combine 0.5 ml sample with 0.5 ml electrophoresis sample buffer in screw capped test tube. Apply caps, but keep loose.

5. Heat capped tubes for 3 min in boiling water.

6. Using the volumes based on protein content calculations, and considering the fact that each extract has been diluted 1 : 1 with sample preparation buffer, with a syringe apply 10 and 20 μg protein of each fish extract to wells of the stacking gel.

7. Apply 10 μl of molecular weight standards to one sample well.

8. Follow manufacturer's instructions to assemble and run the electrophoresis unit. Shut off the power supply when the line of Bromophenol Blue tracking dye has reached the bottom of the separating gel. Disassemble the electrophoresis unit and carefully remove the separating gel from between the plates. Place the gel in a flat dish with the Coomassie Brilliant Blue Stain Solution. Allow the gel to stain for at least 30 min. (If possible, place the dish with the gel on a gentle shaker during staining and destaining.) Pour off the stain solution, then destain the gel for at least 2 hr using the Destain Solution, with at least two changes of the solution.

9. Measure the migration distance (cm) from the top of the gel to the center of the protein band for the molecular weight standards and for each of the major protein bands in the fish extract samples. Also measure the migration distance of the Bromophenol Blue tracking dye from the top of the gel.

10. Observe and record the relative intensity of the major protein bands for each fish extract.

DATA AND CALCULATIONS

Protein Determination

Tube Identity	Absorbance	μg protein/50 μl	$\mu g/ml$
Std. 1, 10 μl BSA			
Std. 1, 10 μl BSA			
Std. 2, 20 μl BSA			
Std. 2, 20 μl BSA			
Std. 3, 30 μl BSA			
Std. 3, 30 μl BSA			
Std. 4, 40 μl BSA			
Std. 4, 40 μl BSA			
Std. 5, 50 μl BSA			
Std. 5, 50 μl BSA			
Sample 1 : 5			
Sample 1 : 5			
			$\overline{X} =$
Sample 1 : 10			
Sample 1 : 10			
			$\overline{X} =$
Sample 1 : 20			
Sample 1 : 20			
			$\overline{X} =$

Equation of the line:

Sample calculation for fish extract protein concentration: (for fish extract diluted 1 : 10, and 25 μl analyzed)

Equation of the line: $y = 0.0108x + 0.0022$

If $y = 0.245$, $x = 22.48$

$$(22.48 \ \mu g \ protein/50 \ \mu l) \times (50 \ \mu l/25 \ \mu l)$$

$$\times (10 \ ml/1 \ ml) = 8.99 \ \mu g \ protein/ul$$

How many μl are needed to get 20 μg protein?

$$(8.99 \ \mu g \ protein/\mu l) \times Z \ \mu l = 20 \ \mu g$$

$$Z = 2.22 \ \mu l$$

Because of 1 : 1 dilution with sample preparation buffer, use 4.44 μl to get 20 μg protein.

Electrophoresis

1. Calculate the relative mobility of three major protein bands and all the molecular weight standards. To determine the relative mobility (R_f) of a protein, divide its migration distance from the top of the gel to the center of the protein band by the migration distance of the Bromphenol Blue tracking dye from the top of the gel.

$$R_f = \frac{\text{distance of protein migration}}{\text{distance of tracking dye migration}}$$

Sample Identity	Distance of Protein Migration	Distance of Tracking Dye Migration	Relative Mobility	Molecular Weight
Molecular weight standards				
1				
2				
3				
4				
5				
Fish species				
Freshwater				
Saltwater				

2. Prepare a standard curve by plotting relative mobility (x-axis) versus log molecular weight of standards (y-axis).

3. Using the standard curve, estimate the molecular weight of the major proteins in the freshwater and saltwater fish extracts.

QUESTIONS

1. Describe how you would prepare 1 L of the buffer used to extract the fish muscle proteins (0.15 M sodium chloride, 0.05 M sodium phosphate, pH 7.0). Show all calculations.

2. Discuss the differences between the fish species regarding which proteins are present and the concentration of these proteins.

REFERENCES

Chang, S.K.C. 2003. Protein analysis. Ch. 9, in *Food Analysis*, 3rd ed. S.S. Nielsen (Ed.), Kluwer Academic, New York.

Etienne, M. (and 16 other authors). 2000. Identification of fish species after cooking by SDS-PAGE and urea IEF: a collaborative study. *Journal of Agricultural and Food Chemistry* 48: 2653–2658.

Etienne, M. (and 7 other authors). 2001. Species identification of formed fishery products and high pressure-treated fish by electrophoresis: A collaborative study. *Food Chemistry* 72: 105–112.

Laemmli, U.K. 1970. Cleavage of structural proteins during the assembly of the head of bacteriophage T4. *Nature* 227: 680–685.

Pierce. 2001. *Instructions: Micro BCA Protein Assay Reagent Kit*. Pierce, Rockford, IL.

Piñeiro, C. (and 12 other authors). 1999. Development of a sodium dodecyl sulfate–polyacrylamide gel electrophoresis reference method for the analysis and identification of fish species in raw and heat-processed samples: A collaborative study. *Electrophoresis* 20: 1425–1432.

Smith, D.M. 2003. Protein separation and characterization. Ch. 15, in *Food Analysis*, 3rd ed. S.S. Nielsen (Ed.), Kluwer Academic, New York.

Smith, P.K. (and 9 other authors). 1985. Measurement of protein using bicinchoninic acid. *Analytical Biochemistry* 150: 76–85.

Enzyme Analysis to Determine Glucose Content

Laboratory Developed by
Dr Charles Carpenter,
Utah State University, Department of Nutrition and Food Sciences, Logan, Utah

INTRODUCTION

Background

Enzyme analysis is used for many purposes in food science and technology. Enzyme activity is used to indicate adequate processing, to assess enzyme preparations, and to measure constituents of foods that are enzyme substrates. In this experiment, the glucose content of corn syrup solids is determined using the enzymes glucose oxidase and peroxidase. Glucose oxidase catalyzes the oxidation of glucose to form hydrogen peroxide (H_2O_2), which then reacts with a dye in the presence of peroxidase to give a stable colored product. Some kinetic parameters for glucose oxidase also will be determined in this experiment.

This experiment as described uses individual commercially available reagents, but enzyme test kits that include all reagents as a package also are available to quantitate glucose. Such enzyme test kits are available to quantitate various components of foods and other materials. Companies that sell enzyme test kits usually provide detailed instructions for the use of these kits, including information about the following: (1) principle of the assay, (2) contents of the test kit, (3) preparation of solutions, (4) stability of solutions, (5) procedure to follow, (6) calculations, and (7) further instructions regarding dilutions and recommendations for specific food samples.

Reading Assignment

BeMiller, J.N. 2003. Carbohydrate analysis. Ch. 10, in *Food Analysis*, 3rd ed. S.S. Nielsen (Ed.), Kluwer Academic, New York.

Powers, J.R. 2003. Application of enzymes in food analysis. Ch. 16, in *Food Analysis*, 3rd ed. S.S. Nielsen (Ed.), Kluwer Academic, New York.

Objective

Determine the glucose content of food products using the enzymes glucose oxidase and peroxidase.

Principle of Method

Glucose is oxidized by glucose oxidase to form hydrogen peroxide, which then reacts with a dye in the presence of peroxidase to give a stable colored product that can be quantitated spectrophotometrically (coupled reaction).

Chemicals

	CAS No.	Hazards
Acetic acid (CH_3COOH)	64-19-7	Corrosive
o-Dianisidine·2HCl	20325-40-0	Tumor causing, carcinogenic
D-Glucose	50-99-7	
Glucose oxidase	9001-37-0	
Horseradish peroxidase	9003-99-0	
Sodium acetate (CH_3COONa)	127-09-3	
Sulfuric acid (H_2SO_4)	7664-93-9	Corrosive

Reagents

- Acetate buffer, 0.1 M, pH 5.5
  Dissolve 13.61 g sodium acetate in ca. 800 ml water in a 1-L beaker. Adjust pH to 5.5 using concentrated acetic acid. Dilute to 1 liter in a volumetric flask.
- Glucose test solution
  In a 100-ml volumetric flask, dissolve 0.4 ml glucose oxidase (1000 units/ml), 40 mg horseradish peroxidase, and 40 mg o-dianisidine · 2HCl in the 0.1 M acetate buffer. Dilute to volume with the acetate buffer and filter as necessary.
- Glucose standard solution, 1 mg/ml
  Accurately weigh ca. 1 g glucose (record exact weight), transfer to a 1-L volumetric flask, and dilute to volume with water. Mix and let stand for 2 hr to let mutorotation occur.
- Sulfuric acid, diluted (1 part H_2SO_4 + 3 parts water)
  In a 1-L beaker in the hood, add 450 ml water, then add 150 ml H_2SO_4. This will generate a lot of heat.

Hazards, Precautions, and Waste Disposal

Concentrated sulfuric acid is an extreme corrosive; avoid contact with skin and clothes and breathing vapors. Acetic acid is corrosive and flammable. Wear safety glasses at all times and corrosive resistant gloves. Otherwise, adhere to normal laboratory safety procedures. The o-dianisidine · 2HCl must be disposed of as a hazardous waste. Other waste likely may be put down the drain using a water rinse, but follow good laboratory practices outlined by environmental health and safety protocols at your institution.

Supplies

- Beaker, 1 L
- Corn syrup solids (or high fructose corn syrup), 0.25 g
- Mechanical, adjustable volume pipettors, 1000 μl and 200 μl with plastic tips
- 5 Spatulas
- 20 Test tubes, 18 × 150 mm, heavy-walled to keep from floating in water bath
- Test tube rack
- 3 Volumetric flasks, 100 ml
- Volumetric flask, 250 ml
- 2 Volumetric flasks, 1 L
- Weighing paper

Equipment

- Analytical balance
- Mechanical, adjustable volume pipettor, 1000 μl, with tips
- pH meter
- Spectrophotometer
- Water bath, 30°C

PROCEDURE

(Instructions are given for analysis in triplicate.)

1. Using the glucose standard solution (1 mg/ml) and deionized distilled (dd) water as indicated in the table below, pipette aliquots of the glucose standard into clean test tubes such that the tubes contain 0 to 2 mg of glucose (use 1000 μl mechanical pipettor to pipette samples), in a total volume of 2 ml. These tubes will be used to create a standard curve, with values of 0 to 2 mg glucose/2 ml. The 0 mg glucose/2 ml sample will be used to prepare the reagent blank. Do this in duplicate.

| | mg Glucose/2 ml | | | | |
	0	0.10	0.15	0.25	0.3
ml glucose std solution	0	0.10	0.15	0.25	0.3
ml dd water	2.0	1.90	1.85	1.75	1.7

2. Accurately weigh ca. 0.25 g corn syrup solids. Dilute with water to volume in a 250-ml volumetric flask (Sample A).
3. Using volumetric flasks, make two further dilutions of Sample A: (a) 10-ml Sample A diluted to 100 ml with water (Sample B), and (b) 1-ml Sample A diluted to 100 ml with water (Sample C).

4. In triplicate, add 2 ml of each dilution to test tubes. This will let you determine glucose concentrations in samples ranging from 1 to 100% glucose.
5. Put all tubes in water bath at 30°C for 5 min. Add 1.0 ml glucose test solution to each tube at 30-sec intervals.
6. After exactly 30 min, stop the reactions by adding 10 ml of the diluted H_2SO_4.
7. Cool to room temp. Measure absorbance at 540 nm against reagent blank.

DATA AND CALCULATIONS

Absorbance of standard solutions:

| | mg Glucose/2 ml | | | |
Trial	0	0.10	0.15	0.25
1				
2				

Absorbance of samples:

Trial	Sample A	Sample B	Sample C
1			
2			
3			

Glucose content of samples:

| | g Glucose/2 ml (from Standard Curve) | | | Calculated % Glucose | | |
Trial	Sample A	Sample B	Sample C	Sample A	Sample B	Sample C
1						
2						
3						
	$\overline{X} =$ SD =	$\overline{X} =$ SD =	$\overline{X} =$ SD =			

1. Plot absorbance on the y-axis versus mg glucose/2 ml on the x-axis.
2. Determine the equation of the line. This is the standard curve of this reaction.
3. Determine the concentration of glucose in the diluted samples (Samples A, B, and C).

Sample calculation:

Equation of the line: $y = 0.368x = 0.05$

If $y = 0.4$, $x = 0.951$

$$(0.951 \text{ mg glucose}/2 \text{ ml}) \times (100 \text{ ml}/10 \text{ ml})$$
$$\times (250 \text{ ml}/0.25 \text{ g}) = 4755 \text{ mg/g}$$
$$= 4.755 \text{ g glucose/g}$$

Questions

1. Explain why this experiment is said to involve a coupled reaction. Write in words the equations for the reactions. What conditions must be in place to ensure accurate results for such a coupled reaction?
2. How do the results obtained compare to specifications for the commercial product analyzed?

For questions 3-5, assume this is a single reaction, and not a coupled reaction:

3. How would you obtain experimentally (in the laboratory) the velocity of reaction (V_i) for this enzyme?
4. How would you plot V_i versus concentration in order to obtain several points, which would permit you draw a Michaelis–Menten saturation curve?
5. What would this plot look like (draw it)? Describe mathematically the parameters that can be obtained from this equation and their importance.

REFERENCES

BeMiller, J.N. 2003. Carbohydrate analysis. Ch. 10, in *Food Analysis*, 3rd ed. S.S. Nielsen (Ed.), Kluwer Academic, New York.

Powers, J.R. 2003. Application of enzymes in food analysis. Ch. 16, in *Food Analysis*, 3rd ed. S.S. Nielsen (Ed.), Kluwer Academic, New York.

Gliadin Detection in Food by Immunoassay

Laboratory Developed by
Mr Gordon Grant and Dr Peter Sporns
University of Alberta, Deparment of Agricultural, Food, and Nutritional Science, Edmonton, Alberta, Canada

INTRODUCTION

Background

Immunoassays are very sensitive and efficient tests that are commonly used to identify a specific protein. Examples of applications in the food industry include identification of proteins expressed in genetically modified foods, drug residues, or proteins associated with a disease, including celiac disease. This genetic disease is associated with Europeans and affects about one in every 200 people in North America. These individuals react immunologically to wheat proteins, and consequently their own immune systems attack and damage their intestines. This disease is managed if wheat proteins, specifically "gliadins," are avoided in foods.

Wheat protein makes up 7 to 15% of a wheat grain. About 40% of the wheat proteins are various forms of gliadin protein, which are also found in oat, barley, rye and other grain flours and related starch derivatives. Gliadin proteins are described as prolamins, a protein classified based on its extractability with aqueous alcohol. Rice and corn are two common grains that do not contain significant gliadin protein, and are well tolerated by those with celiac disease.

The immune system of all animals can respond to many foreign substances by the development of specific antibodies. Antibody proteins bind strongly to and assist in the removal of a foreign substance in the body. Animals make antibodies against many different "antigens," defined as any substance that will elicit a specific immune response in the host. These include foreign proteins, peptides, carbohydrates, nucleic acids, lipids, and many other naturally occurring or synthetic compounds.

Immunoassays are tests that take advantage of the remarkably specific and strong binding of antibodies to antigens. Antibodies that identify a specific protein can be developed by injection of a laboratory animal with this protein, much as humans are vaccinated against a disease. These antigen-specific antibodies can be used to identify the antigen in a food through appropriate use of enzyme markers linked covalently to either the antibody or a reference antigen. This is why the general name for these types of assays is "Enzyme-Linked Immunosorbent Assay" or ELISA. The "sorbent" part of ELISA infers there is an adsorbing solid phase to which antibodies ("immuno" part of ELISA) or antigens are fixed non-specifically.

ELISAs can be used to determine the presence and quantity of either antibody or antigen and are readily automated. They can be scaled up such that many hundreds of samples can be processed and analyzed daily. Also, this type of immunoassay concept can be used to determine if individuals have celiac disease by analyzing for the presence of gliadin-specific antibodies in their blood.

Reading Assignment

Sporns, P. 2003. Immunoassays. Ch. 17, in *Food Analysis*, 3rd ed. S.S. Nielsen (Ed.), Kluwer Academic, New York.

Objective

Determine the presence of gliadin in various food products using a rabbit anti-gliadin antibody horseradish peroxidase conjugate in a dot blot nitrocellulose ELISA.

Principle of Method

A dot blot ELISA will be used in this lab to detect gliadin in food samples. Dot blot assays use nitrocellulose paper for a solid phase. Initially gliadin proteins are purified by differential centrifugation, in which most of the non-gliadin proteins are washed away with water and NaCl solutions, then the gliadin is extracted with a detergent solution. A drop of the sample or standard antigen is applied to the nitrocellulose paper, where it adheres non-specifically. The remaining non-specific binding sites are then "blocked" using a protein unrelated to gliadin, namely bovine serum albumin. The bound gliadin antigen in the food spot then can be reacted with an antigen-specific antibody-enzyme conjugate. Theoretically, this antibody probe then will bind only to gliadin antigen bound already to the nitrocellulose paper. Next, the strip is washed free of unbound antibody-enzyme conjugate, then placed in a substrate solution in which an enzymatically catalyzed precipitation reaction can occur. Brown colored "dots" indicate the presence of gliadin-specific antibody, and hence gliadin antigen.

Chemicals

	CAS No.	Hazards
Bovine serum albumin (BSA)	9048-46-8	
Chicken egg albumin (CEA)	9006-59-1	
3,3'-Diaminobenzidine tetrahydrochloride (DAB)	7411-49-6	
Hydrogen peroxide, 30% (H_2O_2)	7722-84-1	Oxidizing, corrosive
Rabbit anti-gliadin immunoglobin conjugated to horseradish peroxidase (RAGIg-HRP Sigma A1052)		
Sodium chloride (NaCl)	7647-14-5	Irritant
Sodium dodecyl sulfate (SDS)	151-21-3	Harmful
Sodium phosphate, monobasic ($NaH_2PO_4 \cdot H_2O$)	7558-80-7	Irritant
Tris(hydroxymethyl) aminomethane (TRIS)	77-86-1	Irritant
Tween-20 detergent	9005-64-5	

Reagents

(**It is recommended that these solutions be prepared by the laboratory assistant before class.)

- Blocking solution**
    3% BSA in PBS; 5–10 ml per student
- DAB substrate**
    60 mg DAB dissolved in 100 ml 50 mM TRIS pH 7.6, then filtered through Whatman #1. (*Note*: DAB may not completely dissolve in this buffer if it is the free base form instead of the acid form. Just filter out the undissolved DAB and it will still work well.) Just 5 min prior to use, add 100 μl 30% H_2O_2; 5–10 ml per student.
- Gliadin antibody probe**
    1/500 RAGlg-HRP + 0.5% BSA in PBST; 5–10 ml per student.
- Gliadin extraction detergent
    1% SDS in water; 10 ml per student.
- Gliadin standard protein, 4000 μg/ml, in 1% SDS**
    One vial of 150 μl per student.
- Negative control sample
    3% CEA (or other non-gliadin protein) in PBST; 0.1 ml per student.
- Phosphate-buffered saline (PBS)**
    0.05 M sodium phosphate, 0.9% NaCl, pH 7.2; for dissolving negative control sample.
- Phosphate-buffered saline + Tween 20 (PBST)**
    0.05 M sodium phosphate, 0.9 % NaCl, 0.05% Tween 20, pH 7.2; 250 ml per student.

Hazards, Precautions, and Waste Disposal

Adhere to normal laboratory safety procedures. Wear gloves and safety glasses at all times. Handle the DAB substrate with care. Wipe up spills and wash hands thoroughly. The DAB, SDS, and hydrogen peroxide wastes should be disposed of as hazardous wastes. Other wastes likely may be put down the drain using a water rinse, but follow good laboratory practices outlined by environmental health and safety protocols at your institution.

Supplies

(Used by students)

- 1–10 μl, 10–100 μl, and 200–1000 μl positive displacement pipettors
- Disposable tips, for pipettors
- Filter paper, Whatman #1
- Food samples (e.g., flour, crackers, cookies, starch, pharmaceuticals, etc.)
- Funnels, tapered glass
- Mechanical, adjustable volume pipettors, for 2 μl, 100 μl, and 1000 μl ranges, with plastic tips (glass capillary pipets can be substituted for 2 μl pipettors)
- Microcentrifuge tubes, 1.5 ml, 2 per sample processed
- Nitrocellulose paper (BioRad 162-0145) cut into 1.7 cm × 2.3 cm rectangular strips (NC strips)
- Petri dishes, 3.5 cm
- Test tubes, 13 × 100 mm, six per student
- Test tube rack, one per student
- Tissue paper
- Tweezers, one set per student
- Wash bottles, one per two students

Equipment

- Mechanical platform shaker
- Microcentrifuge
- pH meter
- Vortex mixer

PROCEDURE

Sample Preparation

(*Note*: Sample preparation by the students may take place on a separate day prior to the ELISA. In this initial sample preparation lab, the principles of differential centrifugation, with respect to the Osborne Protein Classification system, may be studied. Sample preparation and immunoassay may not be reasonable to achieve in one day. If only one day can be allocated for this lab, the samples can be prepared ahead of time for the students by the technical assistant, and the purpose of this lab will focus solely on immunoassay concepts and technique.)

1. Weigh accurately (record the mass), about 0.1 g of flour, starch, or a ground processed food and add to a 1.5-ml microcentrifuge tube. Add 1.0 ml distilled water and vortex for 2 min. Place in the microcentrifuge with other samples and centrifuge at 800 × g for 5 min. Discard the supernatant (albumins). Repeat.
2. Add 1.0 ml of 1.5 M NaCl to the pellet from Step 1 and resuspend it by vortexing for 2 min. If the pellet isn't resuspending, dislodge it with a spatula. Centrifuge at 800 × g for 5 min. Discard the supernatant (globulins). Repeat.
3. Add 1.0 ml of 1% SDS detergent to the pellet from Step 2 and resuspend it to extract the gliadins. Vortex for 2 min. Centrifuge at 800 × g or 5 min. Carefully pipette off most of the supernatant and transfer to a clean microcentrifuge tube. Discard the pellet.

Standard Gliadin

The standard pure gliadin is dissolved at a concentration of 4000 μg/ml in 1% SDS. To provide for a series of standards to compare unknown samples, dilute the standard serially by a factor of 10 in 13 × 100 mm test tubes to make 400 μg/ml, 40 μg/ml, and 4 μg/ml standards in 1% SDS. Use 100 μl of the highest standard transferred to 900 μl of 1% SDS detergent for the first 10-fold dilution. Repeat this procedure serially to produce the last two standards. As each standard is made, mix it well on a Vortex mixer.

Nitrocellulose Dot Blot ELISA

[*Note*: The nitrocellulose (NC) strips should only be handled with tweezers to prevent binding of proteins and other compounds from your fingers. Hold the strips with the tips of the tweezers on the corners of the nitrocellulose to avoid damaging or interfering with the spotted surface.]

1. Mark 2 NC strips with a pencil into 6 equal boxes each (see drawing below).
2. Pipette 2 μl each of sample, standard or negative controls onto the nitrocellulose paper. Lay the nitrocellulose strips flat onto some tissue.
   a. On nitrocellulose strip A, pipette 2 μl of 4 different SDS food sample extracts.
   b. On nitrocellulose strip B, pipette 2 μl of the 4 different gliadin standards.
   c. On the remaining two squares (5 and 6) on strips A and B, add 2 μl of 1% SDS and 2 μl of the protein 3% BSA control, respectively.
   d. Let the spots air dry on the tissue.

Below are diagrams of the NC strips marked off in boxes and numbered by pencil. The circles are **not** penciled in, but rather they represent where the 2 μl sample or standard spots will be applied.

A series (food sample):

1 = food sample 1@1× dilution
2 = food sample 2@1× dilution
3 = food sample 3@1× dilution
4 = food sample 4@1× dilution
5 = 1% SDS control
6 = 3% protein negative control (CEA)

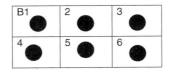

B series (gliadin standards):

1 = gliadin standard@4000 μg/ml
2 = gliadin standard@400 μg/ml
3 = gliadin standard@40 μg/ml
4 = gliadin standard@4 μg/ml
5 = 1% SDS control
6 = 3% protein negative control (CEA)

3. Place both NC strips into a petri dish containing 5 ml of blocking solution (3% BSA in PBS) and let incubate 20 min on the mechanical shaker so that the NC strips are moving around slightly in the solution.
4. Rinse the strips well with PBST using a wash bottle over a sink, holding the strips by the tips of their corners with tweezers.
5. Place the nitrocellulose strips into a petri dish containing about 5 ml of 1/500 RAGIg-HRP conjugate, and incubate for 60 min on the mechanical shaker.
6. Wash the nitrocellulose strips with PBST using a wash bottle, then incubate them for 5 min in a clean petri dish half full with PBST. Rinse again well with PBST, and rinse one last time with distilled water.
7. Add NC strips to a petri dish containing the 5 ml of DAB/H$_2$O$_2$ substrate and watch for the development of a brown stain. (*Note*: Handle the substrate with care. Wipe up spills, wash hands thoroughly, and wear gloves. Although there is no specific evidence that DAB is carcinogenic like the related benzidine compound, it should be treated as if it were.) Stop the reaction in 10–15 min, or when the background nitrocellulose color is becoming noticeably brown, by rinsing each strip in distilled water.
8. Let the NC strips air dry on tissue paper.

DATA AND CALCULATIONS

Make any observations you feel are pertinent to this laboratory. Attach the developed NC strips to your lab report with a transparent tape.

Describe the results based on observations of the degree of brown colored stain in standards and samples relative to negative controls. You can use a crude quantitative rating system like + + +, + +, +, +/−, − to describe and report the relative intensities of the dot reactions (*Note*: the brown dot images will fade in several days).

Make very crude approximations of the quantity of gliadin in each substance relative to (more or less than) the standard gliadin dots. Comment on this crude estimate relative to the food product's gluten status [i.e., gluten free or not; *Codex alimentarius* (http://www.codexalimentarius.net/) defines less than 40 mg of gliadins/100 g of food to be classified as "gluten-free"].

Tabulate your results in a manner that is easy to interpret.

To make the gluten status estimation, you must know the values of both the concentration of the food sample (g food/ml extraction solution) extracted and the concentration of the gliadin standards (mg gliadin/ml extraction solution) to which you are making a comparison.

Example calculation:

If the food sample has a concentration of 100 mg/ml and reacts equivalently to a 4 μg/ml gliadin standard, it can be estimated that 4 μg gliadin is in 100 mg food sample, because both are applied at equal volumes so the two concentrations can be related fractionally (i.e., 4 μg gliadin/100 mg food sample). Since this sample has a gliadin concentration less than the limit set by *Codex Alimentarius* (40 μg gliadin/100 mg or 40 mg gliadins/100 g food), the food can be considered "gliadin free").

QUESTIONS

1. Draw a set of symbolic pictures representing the stages of the dot blot assay used in this laboratory, including the major active molecular substances being employed (i.e., nitrocellulose solid matrix, antigen, BSA blocking reagent, antibody-enzyme conjugate, antigen, substrate, product).
2. Why should you block unbound sites on nitrocellulose with 3% BSA in a special blocking step after applying samples to the membrane?
3. Why is a protein negative control spot (BSA) used in the dot blot?
4. Describe the basic role of horseradish peroxidase enzyme (i.e., why is it attached to the rabbit antibody), and what roles do 3,3'-diaminobenzidine and H_2O_2 play in the development of the colored dot reaction in this ELISA? Do not describe the actual chemical reaction mechanisms, but rather explain why a color reaction can ultimately infer a gliadin antigen is present on the nitrocellulose paper.

REFERENCES

Miletic, I.D., Miletic, V.D., Sattely-Miller, E.A., and Schiffman, S.S. 1994. Identification of gliadin presence in pharmaceutical products. *Journal of Pediatric Gastroenterology and Nutrition* 19: 27–33.

Sdepanian, V.L, Scaletsky, I.C.A., Fagundes-Neto, U., and deMorais, M.B. 2001. Assessment of gliadin in supposedly gluten-free foods prepared and purchased by celiac patients. *Journal of Pediatric Gastroenterology and Nutrition* 32: 65–70.

Skerritt, J.H., and Hill, A.S. 1991. Enzyme immunoassay for determination of gluten in foods: Collaborative study. *Journal of the Association of Official Analytical Chemists* 74: 257–264.

Sporns, P. 2003. Immunoassays. Ch. 17, in *Food Analysis*, 3rd ed. S.S. Nielsen (Ed.), Kluwer Academic, New York.

16 chapter

Examination of Foods for Extraneous Materials

INTRODUCTION

Background

Extraneous materials are any foreign substances in foods that are associated with objectionable conditions or practices in production, storage, or distribution of foods. Extraneous materials include: (a) filth or objectionable matter contributed by animal contamination (rodent, insect, or bird matter) or unsanitary conditions; (b) decomposed material or decayed tissues due to parasitic or nonparasitic causes; and (c) miscellaneous matter (sand, soil, glass, rust, or other foreign substances). Bacterial contamination is excluded from these substances.

Filth is classified according to its extractability. Light filth is oleophilic and lighter than water (separated from product by floating it in an oil-aqueous mixture). Insect fragments, rodent hairs, and feather barbules are examples of light filth. Heavy filth is heavier than water and separated from the product by sedimentation based on different densities of filth, food particles, and immersion liquids ($CHCl_3$, CCl_4, etc.). Examples of heavy filth are sand, soil, and nutshell fragments. Sieved filth involves particles separated from the product by use of selected mesh sizes. Whole insects, stones, sticks, and bolts are examples of sieved filth.

Various methods of isolation of extraneous matter from various food commodities can be found in the *Official Methods of Analysis* of the AOAC International and in the *Official Methods* of the American Association of Cereal Chemists (AACC). Presented here are a few procedures for some food commodities, with descriptions based on Association of Official Analytical Chemists (AOAC) methods, but the quantities reduced to half.

Reading Assignment

Pedersen, J. 2003. Extraneous matter. Ch. 20, in *Food Analysis*, 3rd ed. S.S. Nielsen (Ed.), Kluwer Academic, New York.

Notes

Regulatory examination of samples by the Food and Drug Administration (FDA) would be based on replicate samples using official methods, including the specified sample size. However, for instructional purposes, the costs associated with adequate commercial 1-L Wildman trap flasks, reagents, and food samples specified in official methods may be prohibitive. Procedures given below are based on AOAC methods, but all quantities are reduced to half, and a 500-ml Wildman trap flask (vs. 1-L trap flask) is specifed in most procedures. Commercially available 1-L trap flasks with the standard stopper rod would ideally be used (with all quantities in the procedures doubled). However, 500-ml

trap flasks can be made for use in this experiment. To do this, drill a hole through a rubber stopper of a size just too large for a 500-ml Erlenmeyer flask. Thread a heavy string through the hole in the stopper, and knot both ends of the string. Coat the sides of the rubber stopper with glycerin and *carefully* force it (with larger end of stopper pointed up) through the top of the flask. Note that the string could be a trap for contaminants such as rodent hair and insect fragments.

For the parts of this laboratory exercise that require filter paper, S&S #8 (Schleicher & Schuell, Inc., Keene, NH) is recommended. It meets the specifications set forth in AOAC Method 945.75 Extraneous Materials (Foreign Matter) in Products, Isolation Techniques Part B(i), which suggested using "smooth, high wet strength, rapid acting filter paper ruled with oil-, alcohol-, and water-proof lines 5 mm apart." The S&S #8 ruled filter paper is 9 cm in diameter, and fits well into the *top* of standard 9-cm *plastic* petri dishes. The *bottom* of the plastic petri dish can be used as a protective cover over the sample filter paper in the top of the petri dish. The top of the plastic petri dish provides a 9-cm flat surface (as opposed to glass petri dishes) for examining the filter paper, making it easier to view the plate without having to continuously refocus the microscope. The 5-mm ruled lines provide a guide for systematically examining and enumerating contaminants on the filter paper at 30× magnification. To obtain a moist surface on which contaminants can be manipulated and observed, apply a small amount of glycerin: 60% alcohol (1 : 1) solution to the top of the petri dish before transferring the filter paper from the Buchner funnel. Using both overhead and sub-stage lighting with the microscope will assist in identifying contaminants.

Objective

The objective of this laboratory is to utilize techniques to isolate the extraneous matter from various foods: cottage cheese, jam, infant food, potato chips, and citrus juice.

Principle of Methods

Extraneous materials can be separated from food products by particle size, sedimentation, and affinity for oleophilic solutions. Once isolated, extraneous materials can be examined microscopically.

METHOD A: EXTRANEOUS MATTER IN SOFT CHEESE

Chemicals

	CAS No.	Hazards
Phosphoric acid (H_3PO_4)	7664-38-2	Corrosive

Reagents

- Phosphoric acid solution, 400–500 ml
  Combine 1 part phosphoric acid with 40 parts deionized distilled (dd) water (vol/vol).

Hazards, Precautions, and Waste Disposal

Adhere to normal laboratory safety procedures. Wear safety glasses at all times. Waste likely may be put down the drain using a water rinse, but follow good laboratory practices outlined by environmental health and safety protocols at your institution.

Supplies

- Beaker, 1 L (for phosphoric acid solution)
- Beaker, 600 ml (to boil water)
- Buchner funnel
- Cottage cheese, 115 g
- Filter paper
- Heavy gloves
- Pipette, 10 ml (to prepare phosphoric acid solution)
- Pipette bulb or pump
- Spoon
- Side-arm flask, 500 ml or 1 L
- Stirring rod
- Tap water, ca. 500 ml (boiling)
- Tweezers
- Volumetric flask, 500 ml (to prepare phosphoric acid solution)
- Weighing boat

Equipment

- Hot plate
- Microscope
- Top loading balance
- Water aspirator system

Procedure

(Based on AOAC Method 960.49, Filth in Dairy Products.)

1. Weigh out 115 g cottage cheese and add it to 400–500 ml boiling phosphoric acid solution (1+40 mixture) in a 1-L beaker, stirring with a glass stirring rod continuously to disperse the cottage cheese.
2. Filter the mixture through filter paper in a Buchner funnel, using a vacuum created by a water aspirator. Do not let the mixture accumulate on the paper, and continually wash filter with a stream of hot water to prevent clogging. Make sure the cheese mixture is hot as it is filtered. When filtration is impeded, add hot water or phosphoric acid solution (1 + 40 mixture) until the paper clears. [May also use dilute (1–5%) alkali or hot alcohol to aid in filtration.] Resume addition of sample and water until sample is filtered.
3. Examine filter paper microscopically.

METHOD B: EXTRANEOUS MATTER IN JAM

Chemicals

	CAS No.	Hazards
Heptane (12.5 ml)	142-82-5	Harmful, highly flammable, dangerous for the environment
Hydrochloric acid, concentrated (HCl) (5 ml)	7647-01-0	Corrosive

Hazards, Precautions, and Waste Disposal

Heptane is an extremely flammable liquid; avoid open flames, breathing vapors and contact with skin. Otherwise, adhere to normal laboratory safety procedures. Wear safety glasses at all times. Dispose of heptane waste as a hazardous waste. Other waste may be put down the drain using a water rinse.

Supplies

- 2 Beakers, 250 ml (for weighing jam and heating water)
- Buchner funnel
- Filter paper
- Glass stirring rod
- Graduated cylinder, 100 ml
- Ice water bath (to cool mixture to room temperature)
- Jam, 50 g
- Measuring pipette, 10 ml (for heptane)
- Pipette bulb or pump
- Side-arm flask, 500 ml or 1 L
- Spoon
- Thermometer
- Tweezers
- Volumetric pipette, 5 ml (for conc. HCl)
- Waste jar (for heptane)
- Water, dd, 100 ml (heated to 50°C)
- Wildman trap flask, 500 ml

Equipment

- Hot plate
- Microscope
- Top loading balance
- Water aspirator system

Procedure

(Based on AOAC Method 950.89, Filth in Jam and Jelly.)

1. Empty contents of jam jar into beaker and mix thoroughly with glass stirring rod.
2. Weigh 50 g of jam into a beaker, add ca. 80 ml dd water at 50°C, transfer to a 500-ml trap flask, (use the other ca. 20 ml dd water to help make transfer), add 5 ml conc. HCl, and boil for 5 min.
3. Cool to room temperature (with an ice water bath).
4. Add 12.5 ml heptane and stir thoroughly.
5. Add dd water to a level so heptane is just above rubber stopper when in the "trap" position.
6. Trap off the heptane, and filter the heptane through filter paper in a Buchner funnel, using vacuum created by a water aspirator.
7. Examine filter paper microscopically.

METHOD C: EXTRANEOUS MATTER IN INFANT FOOD

Chemicals

	CAS No.	Hazards
Light mineral oil (10 ml)	8012-95-1	

Hazards, Precautions, and Waste Disposal

Adhere to normal laboratory safety procedures. Wear safety glasses at all times. Waste may be put down the drain using water rinse.

Supplies

- Baby food, ~113 g (1 jar)
- Buchner funnel
- Filter paper
- Glass stirring rod
- Graduated cylinder, 10 or 25 ml
- Pipette bulb or pump
- Side-arm flask, 500 ml or 1 L
- Spoon
- Tweezers
- Volumetric pipette, 10 ml

- Water, deaerated, 500 ml
- Wildman trap flask, 500 ml

Equipment

- Microscope
- Water aspirator system

Procedure

(Based on AOAC Method 970.73, Filth in Pureed Infant Food, A. Light Filth.)

1. Transfer 113 g (1 jar) of baby food to a 500-ml trap flask.
2. Add 10 ml of light mineral oil, and mix thoroughly.
3. Fill the trap flask with deaerated water (can use dd water) at room temperature.
4. Let stand 30 min, stirring 4–6 times during this period.
5. Trap off mineral oil in a layer above the rubber stopper, then filter the mineral oil through filter paper in a Buchner funnel, using vacuum created by a water aspirator.
6. Examine filter paper microscopically.

METHOD D: EXTRANEOUS MATTER IN POTATO CHIPS

Chemicals

	CAS No.	Hazards
Ethanol, 95%	64-17-5	Highly flammable
Heptane (9 ml)	142-82-5	Harmful, highly flammable, Dangerous to environment
Petroleum ether (200 ml)	8032-32-4	Harmful, highly flammable, dangerous to environment

Reagents

- Ethanol, 60%, 1 L
  Use 95% ethanol to prepare 1 L of 60% ethanol; dilute 632 ml of 95% ethanol with water to 1 L.

Hazards, Precautions, and Waste Disposal

Petroleum ether, heptane, and ethanol are fire hazards; avoid open flames, breathing vapors, and contact with skin. Otherwise, adhere to normal laboratory safety procedures. Wear safety glasses at all times. Heptane and petroleum ether wastes must be disposed of as hazardous wastes. Other waste may be put down the drain using a water rinse.

Supplies

- Beaker, 400 ml
- Buchner funnel
- Filter paper
- Glass stirring rod
- Graduated cylinder, 1 L (to measure 95% ethanol)
- Ice water bath
- Potato chips, 25 g
- Side-arm flask, 500 ml or 1 L
- Spatula
- Wildman trap flask, 500 ml
- Tweezers
- Volumetric flask, 1 L (to prepare 60% ethanol)
- Waste jars (for heptane and petroleum ether)

Equipment

- Hot plate
- Microscope
- Top loading balance
- Water aspirator system

Procedure

(Based on AOAC Method 955.44, Filth in Potato Chips.)

1. Weigh 25 g of potato chips into a 400-ml beaker.
2. With a spatula or glass stirring rod, crush chips into small pieces.
3. In a hood, add petroleum ether to cover the chips. Let stand 5 min. Decant petroleum ether from the chips through filter paper. Again add petroleum ether to the chips, let stand 5 min and decant through filter paper. Let petroleum ether evaporate from chips in hood.
4. Transfer chips to a 500-ml trap flask, add 125 ml 60% ethanol, and boil for 30 min. Mark initial level of ethanol on flask. During boiling and at end of boiling, replace ethanol lost by evaporation as a result of boiling.
5. Cool in ice water bath.
6. Add 9 ml heptane, mix, and let stand for 5 min.
7. Add enough 60% ethanol to the flask so only the heptane layer is above the rubber stopper. Let stand to allow heptane layer to form at the top, trap off the heptane layer, and filter it through filter paper in a Buchner funnel.
8. Add 9 ml more heptane to solution. Mix, then let stand until heptane layer rises to the top. Trap off the heptane layer, and filter it through filter paper (i.e., new piece of filter paper; not piece used in Parts 3 and 4) in a Buchner funnel.
9. Examine the filter paper microscopically.

METHOD E: EXTRANEOUS MATTER IN CITRUS JUICE

Supplies

- Beaker, 250 ml
- Buchner funnel
- Cheesecloth
- Citrus juice, 125 ml
- Graduated cylinder, 500 ml or 1 liter
- Side-arm flask, 250 ml
- Tweezers

Equipment

- Microscope
- Water aspirator system

Procedure

[Based on AOAC Method 970.72, Filth in Citrus and Pineapple Juice (Canned), Method A. Fly Eggs and Maggots.]

1. Filter 125 ml of juice through a Buchner funnel fitted with a double layer of cheesecloth. Filter with a vacuum created by a water aspirator. Pour the juice slowly to avoid accumulation of excess pulp on the cheesecloth.
2. Examine material on cheesecloth microscopically for fly eggs and maggots.

Questions

1. Summarize the results for each type of food analyzed for extraneous material.
2. Why are contaminants such as insect fragments found in food, when the Pure Food and Drug Act prohibits adulteration?

REFERENCES

AOAC International. 2000. Extraneous matter. Ch. 16, in *Official Methods of Analysis*. AOAC International, Gaithersburg, MD.

Pedersen, J.R. 2003. Extraneous matter. Ch. 20, in *Food Analysis*, 3rd ed. S.S. Nielsen (Ed.), Kluwer Academic, New York.

High Performance Liquid Chromatography

Laboratory Developed in part by
Dr Stephen Talcott,
University of Florida, Department of Food Science, Human Nutrition,
Gainsville, Florida

INTRODUCTION

Background

High performance liquid chromatography (HPLC) has many applications in food chemistry. Food components that have been analyzed with HPLC include organic acids, vitamins, amino acids, sugars, nitrosamines, certain pesticides, metabolites, fatty acids, aflatoxins, pigments, and certain food additives. Unlike gas chromatography, it is not necessary for the compound being analyzed to be volatile. It is necessary, however, for the compounds to have some solubility in the mobile phase. It is important that the solubilized samples for injection be free from all particulate matter, so centrifugation and filtration are common procedures. Also, solid-phase extraction is used commonly in sample preparaton to remove interfering compounds from the sample matrix prior to HPLC analysis.

Many food-related HPLC analyses utilize reversed-phase chromatography in which the mobile phase is relatively polar, such as water, dilute buffer, methanol, or acetonitrile. The stationary phase (column packing) is relatively nonpolar, usually silica particles coated with a C_8 or C_{18} hydrocarbon. As compounds travel through the column they partition between the hydrocarbon stationary phase and the mobile phase. The mobile phase may be constant during the chromatographic separation (i.e., isocratic) or changed stepwise or continuously (i.e., gradient). When the compounds elute separated from each other at the end of the column, they must be detected for identification and quantitation. Identification often is accomplished by comparing the volume of liquid required to elute a compound from a column (expressed as retention volume or retention time) to that of standards chromatographed under the same conditions. Quantitation generally involves comparing the peak height or area of the sample peak of interest to the peak height or area of a standard (at the same retention time). The results are usually expressed in milligrams per gram or milliliters of food sample.

Reading Assignment

Rounds, M.A., and Nielsen, S.S. 2003. Basic principles of chromatography. Ch. 27, in *Food Analysis*, 3rd ed. S.S. Nielsen (Ed.), Kluwer Academic, New York.

Rounds, M.A., and Gregory, J.F. 2003. High performance liquid chromatography. Ch. 28, in *Food Analysis*, 3rd ed. S.S. Nielsen (Ed.), Kluwer Academic, New York.

METHOD A: DETERMINATION OF CAFFEINE IN BEVERAGES BY HPLC

Introduction

The caffeine content of beverages can be determined readily by simple filtration of the beverage prior to separation from other beverage components using reversed-phase HPLC. An isocratic mobile phase generally provides for sufficient separation of the caffeine from other beverage components. However, separation and quantitation is much easier for soft drinks than for a beverage such as coffee, which has many more components. Commercially available caffeine can be used as an external standard to quantitate the caffeine in the beverages by peak height or area.

Objective

To determine the caffeine content of soft drinks by reversed-phase HPLC with ultraviolet (UV) detection, using peak height and area to determine concentrations.

Chemicals

	CAS No.	Hazards
Acetic acid (CH$_3$COOH)	64-19-7	Corrosive
Caffeine	58-08-2	Harmful
Methanol, HPLC grade (CH$_3$OH)	67-56-1	Extremely flammable, toxic

Hazards, Precautions, and Waste Disposal

Adhere to normal laboratory safety procedures. Wear safety glasses at all times. Methanol waste must be handled as a hazardous waste. Other waste likely may be put down the drain using a water rinse, but follow good laboratory practices outlined by environmental health and safety protocols at your institution.

Reagents

(**It is recommended that these solutions be prepared by the laboratory assistant before class.)

- Mobile phase**

  Deionized distilled (dd) water : HPLC-grade methanol : acetic acid, 65 : 35 : 1 (v/v/v), filtered through a Millipore filtration assembly with 0.45 μm nylon membranes and degassed.

- Caffeine solutions of varying concentration for standard curve**

  Prepare a stock solution of 20 mg caffeine/100 ml dd water (0.20 mg/ml). Make standard solutions containing 0.05, 0.10, 0.15, and 0.20 mg/ml, by combining 2.5, 5.0, 7.5, and 10 ml of stock solution with 7.5, 5.0, 2.5, and 0 ml dd water, respectively.

Supplies

(Used by students)

- Disposable plastic syringe, 3 ml (for filtering sample)

- Hamilton glass HPLC syringe, 25 μl (for injecting sample if using manual sample loading)
- Pasteur pipettes and bulb
- Sample vials for autosampler (if using autosampler)
- Soft drinks, with caffeine
- Syringe filter assembly, e.g. Whatman, Cat. #7184001, Membrane filter, Whiteplain, Cellulose nitrate, 13 mm diameter, 0.45 μm pore size (for filtering sample)
- Test tubes, e.g. 13 × 100 mm disposable culture tubes (for filtering sample)

Equipment

- Analytical balance
- HPLC system, with UV–Vis detector

HPLC Conditions

Column	Waters μBondapak C_{18} (Waters, Milford, MA) or equivalent reversed-phase column
Guard column	Waters Guard-Pak Precolumn Module with C_{18} Guard-Pak inserts or equivalent
Mobile phase	dd H_2O : HPLC-grade methanol : acetic acid, 65 : 35 : 1 (v/v/v) (Combine, then filter and degas)
Flow rate	1 ml/min
Sample loop size	10 μl
Detector	Absorbance at 254 nm or 280 nm
Sensitivity	Full scale absorbance = 0.2
Chart speed	1 cm/min

Procedure

(Instructions given for manual injection with strip chart recorder, and for analysis in triplicate.)

1. Filter beverage sample.
   a. Remove plunger from a plastic 3-ml syringe and connect syringe filter assembly (with a membrane in place) to the syringe barrel.
   b. Use a Pasteur pipette to transfer a portion of beverage sample to the syringe barrel. Insert and depress syringe plunger to force sample through the membrane filter and into a small test tube.
2. Flush Hamilton HPLC syringe with filtered sample, then take up 15–20 μl of filtered sample (try to avoid taking up air bubbles).
3. With HPLC injector valve in LOAD position, insert syringe needle into the needle port all the way.
4. Gently depress syringe plunger to completely fill the 10-μl injector loop with sample.

5. Leaving the syringe in position, *simultaneously* turn valve to INJECT position (mobile phase now pushes sample onto the column) and depress chart marker button on detector (to mark start of run on chart recorder paper).
6. Remove syringe. (Leave valve in the INJECT position so that the loop will be continuously flushed with mobile phase, thereby preventing cross contamination.)
7. After caffeine peak has eluted, return valve to LOAD position in preparation for next injection.
8. Identify the chromatogram for your sample by writing your name and the type of sample along the edge of the paper.
9. Repeat Steps 3–7, injecting each caffeine standard solution in duplicate or triplicate. (Note: the laboratory assistant can inject all standard solutions prior to the laboratory session. The peaks from all chromatograms can then be cut and pasted together onto one page to be copied and given to each student.)

Data and Calculations

1. Measure the height (cm) of the caffeine peak for your sample and the caffeine standards.
2. Calculate the area (cm^2) of the caffeine peak for your sample and the caffeine standards: Use the equation for a triangle, area = (width at half-height) (height). (See Chapter 27 in Nielsen, *Food Analysis*.)

Standard curve:

Caffeine Conc. (mg/ml)	Trial	Peak Height (cm)	Peak Area (cm^2)
0.05	1		
	2		
	3		
0.10	1		
	2		
	3		
0.15	1		
	2		
	3		
0.20	1		
	2		
	3		

Samples:

Trial	Peak Height (cm)	Peak Area (cm^2)
1		
2		
3		

3. Construct two standard curves using data from the caffeine standards: (a) Peak height (cm) versus caffeine concentration (mg/ml), and (b) Peak area (cm^2) versus caffeine concentration (mg/ml).
4. Determine the equations of the lines for both standard curves.

Equation of the Line, Based on Peak Height:

Equation of the Line, Based on Peak Area:

5. Calculate the concentration of caffeine in your sample expressed in terms of mg caffeine/ml using (a) the standard curve based on peak height, and (b) the standard curve based on peak area. Report values for each replicate and calculate the means.

Sample caffeine concentration (mg/ml):

Trial	Peak Height	Peak Area
1		
2		
3		
	$\overline{X} =$	$\overline{X} =$
	SD =	SD =

6. Using the mean values determined in Step 5 above, calculate the concentration of caffeine in your sample expressed in terms of milligram of caffeine in a 12-oz. can (1 ml = 0.0338 oz) using (a) the standard curve based on peak height, and (b) the standard curve based on peak area.

Questions

1. Based on the triplicate values and the linearity of your standard curves, which of the two methods used to calculate concentration seemed to work best in this case? Is this what you would have expected, based on the potential sources of error for each method?
2. Why was it important to filter and degas the mobile phase and the samples?
3. How is the "reversed-phase" HPLC used here different from "normal-phase" with regard to stationary and mobile phases, and order of elution?
4. Mobile Phase Composition
   a. If the mobile phase composition was changed from 65:35:1(v/v/v) to 75:25:1 (v/v/v) water:methanol: acetic acid, how would the time of elution (expressed as retention time) for caffeine be changed, and why would it be changed?
   b. What if it was changed from 65:35:1 (water:methanol: acetic acid) to 55:45:1? How would that change the retention time and why?

METHOD B: SOLID-PHASE EXTRACTION AND HPLC ANALYSIS OF ANTHOCYANIDINS FROM FRUITS AND VEGETABLES

Introduction

Anthocyanins are naturally occurring plant pigments known for their diverse colors depending on solution pH. Analysis for anthocyanins is often difficult due to their similar molecular structure and polarity and their diversity of sugar and/or organic acid substituents. Color intensity is a common means of analyzing for anthocyanins since monomeric anthocyanins are colored bright red at low pH values from 1–3 (oxonium or flavylium forms) and are nearly colorless at higher pH values from 4–6 (carbinol or pseudobase forms). A pure anthocyanin in solution generally follows Beer's law, therefore concentration can be estimated from an extinction coefficient when an authentic standard is not available. However, many standards are commercially available with cyanidin 3-glucoside used most often for quantification purposes.

Red-fleshed fruits and vegetables contain many different anthocyanin forms due to their diverse array of esterified sugar substituents and/or acyl-linked organic acids. However, most foods contain up to six anthocyanin aglycones (without sugar or organic acid substituents, referred to as anthocyanidins) that include delphinidin, cyanidin, petunidin, pelargonidin, peonidin, and malvidin (Fig. 17-1). Sample preparation for anthocyanin analysis generally involves solid-phase extraction of these compounds from the food matrix followed by acid hydrolysis to remove sugar and/or organic acid linkages. Anthocyanidins are then easily separated by reversed-phase HPLC.

The use of solid-phase extraction (SPE) is a common chromatographic sample preparation technique used to remove interfering compounds from a biological matrix prior to HPLC analysis. This physical extraction technique is similar to an actual separation on a reversed-phase HPLC column. Although many SPE stationary phases exist, the use of reversed-phase C_{18} is commonly employed for food analysis. On a relative basis, anthocyanins are less polar than other chemical constituents in fruits and vegetables and will readily bind to a reversed-phase C_{18} SPE cartridge. Other compounds such as sugars, organic acids, water-soluble vitamins, or metal ions have little or no affinity to the cartridge. After the removal of these interferences, anthocyanins can then be efficiently eluted with alcohol, thus obtaining a semipurified extract for HPLC analysis.

Separation of compounds by HPLC involves use of a solid support (column) over which a liquid mobile phase flows on a continuous basis. Chemical interactions with an injected sample and the stationary and mobile phases will influence rates of compound elution

17-1 figure   Anthocyanin structures. Common substitutions on the B-ring include: delphinidin (Dp), cyanidin (Cy), petunidin (Pt), pelargonidin (Pg), peonidin (Pn), and malvidin (Mv).

from a column. For compounds with similar polarities, the use of mixtures of mobile phases (gradient elution) is often employed. Reversed-phase stationary phases are most common for anthocyanin separations, and are based on column hydrophobicity of a silica-based column with varying chain lengths of n-alkanes such as C_8 or C_{18}. By setting initial chromatographic conditions to elute with a polar (water) mobile phase followed by an organic (alcohol) mobile phase, anthocyanins will generally elute in order of their polarity.

You will be analyzing anthocyanins isolated from fruits or vegetables for anthocyanidins (aglycones) following SPE and acid hydrolysis to remove sugar glycosides. After sugar hydrolysis, samples will be injected into an HPLC for compound separation. Depending on plant source, you will obtain between 1 and 6 chromatographic peaks representing common anthocyanidins found in edible plants.

Objective

Isolate and quantify anthocyanidin concentration from common fruits and vegetables by reversed-phase HPLC with Vis detection, using spectrophotometric absorbance readings and extinction coefficients of anthocyanidins to determine standard concentrations.

Chemicals

	CAS No.	Hazards
Hydrochloric acid (HCl)	7647-01-0	Corrosive
Methanol (CH_3OH)	67-56-1	Extremely flammable, toxic
o-Phosphoric acid (H_3PO_4)	7664-38-2	Corrosive

Hazards, Precautions, and Waste Disposal

Adhere to normal laboratory safety procedures. Wear safety glasses at all times. Use hydrochloric acid under a fume hood. Methanol waste must be handled as a hazardous waste. Other waste likely may be put down the drain using a water rinse, but follow good laboratory practices outlined by your environmental health and safety protocols.

Reagents

(**It is recommended that these solutions be prepared by a laboratory assistant before class.)

- 4 N HCl in water (for anthocyanin hydrolysis)**
- 0.01% HCl in water (for sample extraction)**
- 0.01% HCl in methanol (for elution from C_{18} cartridge)**
- Mobile Phase A: 100% water (pH 2.4 with o-phosphoric acid)**
- Mobile Phase B: 60% methanol and 40% water (pH 2.4 with o-phosphoric acid)**

(Each mobile phase should be filtered through a 0.45-μm nylon membrane (Millipore) and degassed while stirring using either a nitrogen sparge, under vacuum (ca. 20 min/liter of solvent), or by sonication.)

Supplies

- Beaker, Pyrex, 500 ml (for boiling water for hydrolysis)
- Blender, kitchen-scale, for sample homogenization
- Disposable plastic syringe, 3–5 ml (for filtering sample)
- Filter paper (Whatman #4) and funnels
- Fruit or vegetable that contains anthocyanins (blueberries, grapes, strawberries, red cabbage, blackberries, cherries, or commercial juices that contain anthocyanins)
- Hamilton glass HPLC syringe, 25 μl (for injecting sample)
- Reversed-phase C_{18} cartridge (for SPE, e.g. Waters C_{18} Sep-Pak, WAT051910)

- Syringe filter (0.45 μm PTFE, polytetrafluoroethylene) e.g. Whatman, Cat. #6785–2504
- Test tubes, screw cap, with lids (for anthocyanin hydrolysis)

Equipment

- Analytical balance
- Hot plate
- HPLC system, gradient, with Vis detector (520 nm)
- Membrane filtering device
- Spectrophotometer and cuvettes (1-cm pathlength)

HPLC Conditions

Column	Waters NovaPak C_{18} (WAT044375) or equivalent reversed-phase column.
Guard column	Waters Guard-Pak Precolumn Module with C_{18} Guard-Pak inserts.
Mobile phase	Phase A: 100% water; Phase B: 60% methanol and 40% water (both adjusted to pH 2.4 with o-phosphoric acid)
Flow rate	1 ml/min
Sample loop size	Variable: 10–100 μl
Detector	Visible at 520 nm
Gradient conditions	Linear ramp. Hold time at 100% Phase B after 15 min may vary with column length and/or column packing material.

Time (min)	% Phase A	% Phase B
0	100	0
5	50	50
10	50	50
15	0	100
35	0	100 (end)
37	100	0 (equilibration)

Procedure

I. Sample Extraction

(Note to Instructor: Several different commodities can be evaluated or the experiment replicated as needed.)

1. Weigh ca. 10 g of fruit or vegetable containing anthocyanins (record exact weight) and place in blender. Add ca. 50 ml of water containing 0.01% HCl and blend thoroughly (acidified acetone, methanol, or ethanol are also suitable substitutions for water). Fruit juices that contain anthocyanins can be used without further preparation.
2. Filter homogenate through filter paper and collect aqueous filtrate.
3. Keep refrigerated until needed.

II. Solid-Phase Extraction

1. Condition a reversed-phase SPE cartridge by first washing with 4 ml of 100% methanol followed by 4 ml of water acidified with 0.01% HCl.
2. *Slowly* pass 1–2 ml of juice or filtrate (record exact volume) through the SPE cartridge being careful not to lose visible pigment. Anthocyanins will adhere to the SPE support and less polar compounds such as sugars, organic acids, and ascorbic acid will be removed.
3. Slowly pass an additional 4 ml of water (acidified with 0.01% HCl) through the cartridge to remove residual water-soluble components. Remove residual moisture from the cartridge by pushing air through the cartridge with an empty syringe or by flushing the cartridge with nitrogen gas until dry.
4. Elute anthocyanins with 4 ml of 0.01% HCl in methanol and collect for subsequent hydrolysis.

III. Acid Hydrolysis

[Note to Instructor: It is recommended that a previously extracted sample be acid hydrolyzed before class to save time. A nonhydrolyzed (glycoside) sample can also be analyzed for comparison to hydrolyzed (aglycone) sample.]

1. Pipette 2 ml of anthocyanins, dissolved in methanol, into a screw-cap test tube and add an equal volume of aqueous 4 N HCl (final acid concentration = 2 N) for a 2-fold dilution factor (see calculations below).
2. Under a fume hood, tightly cap the screw-cap vial and place in boiling water for ca. 90 min.
3. Remove test tubes and cool to room temperature before opening the vial. Filter an aliquot through a 0.45 μm PTFE syringe filter for analysis by HPLC.
4. Inject the filtered extract into the HPLC and record peak areas for quantification of each compound (see Fig. 17-2).

Data and Calculations

Authentic standards for select anthocyanins can be obtained from several sources and should be used according to manufacturer's suggestions. If using

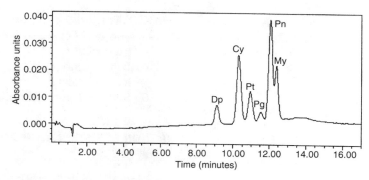

17-2 figure   Typical reversed phase HPLC chromatograph of anthocyanidins (grape).

anthocyanin glycosides, then the acid hydrolysis procedure should be conducted prior to HPLC analysis. Some anthocyanin suppliers include:

1. Cyanidin Chloride (Fisher Scientific, Pittsburgh, PA, Cat.#: A385003M010).
2. Malvidin Chloride [ICN Biomedicals, Costa Mesa, CA, Cat.#: 203888, molecular weight (MW) = 366.75]
3. Various anthocyanin standards (Polyphenolics Laboratories, Sandnes, Norway).
4. Various anthocyanin standards (Indofine Chemical Company, Somerville, NJ).

Cyanidin is a common anthocyanin present in large concentrations in many fruits and vegetables and will be used for example calculations. A standard solution of cyanidin should be prepared in Mobile Phase A (water at pH 2.4) to establish a standard curve. Unless the actual concentration is known from the manufacturer, the standard should be quantified by determining its absorbance on a spectrophotometer at 520 nm against a blank of the same solvent. Using the molar extinction coefficient for cyanidin (obtained from manufacturer; or expressed as cyanidin-3-glucoside equivalents, $\varepsilon = 29,600$ for a 1 M solution and 1-cm light path) the concentration is calculated using Beer's Law: $A = \varepsilon bc$ using the following calculation:

$$\text{mg/L Cyanidin} = \frac{(\text{Absorbance at } \lambda_{max})(1000)(MW)}{\varepsilon}$$

where:

MW ~457 g/mol
$\varepsilon \sim 29,600$

1. Inject a series of standard concentrations into the HPLC to generate a standard curve (*Note*: These procedures can be performed by a laboratory assistant prior to the laboratory session).
2. Create a graph plotting anthocyanin concentration versus peak area to obtain a slope for the

commercial standard and apply to areas of acid hydrolyzed samples (see Fig. 17-3).
3. Express relative concentrations of each identifiable compound as cyanidin equivalents (mg/L) based on their peak area (unless commercial standards are available for each peak in the chromatograph).

Peak	Peak Area	Relative Concentration (mg/L)
1		
2		
3		
4		
5		
6		

mg/L Cyanidin (in unknown)

$$= \frac{\text{Peak Area} \times 2}{\text{Slope of Standard Curve}} \times \text{Sample dilution factors}$$

Peak area multiplied times 2 will compensate for the 2-fold dilution incurred during acid hydrolysis. Sample dilution factors are calculated based on weight of fruit/vegetable per volume of extraction solvent (sample weight + solvent volume/sample weight). Single

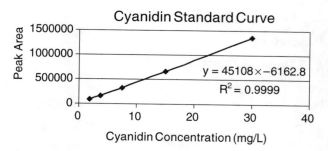

17-3 figure   Typical standard curve for cyanidin.

strength fruit juices would have a sample dilution factor of 1.

Questions

1. Based on chemical structure, why do anthocyanidins elute in their respective order?
2. Predict how each compound would elute from a normal phase column.
3. If the retention time of a compound that had absolutely *no* affinity to the column was 1.5 min and the flow rate was 1 ml/min, what is the total volume of mobile phase contained in the column, tubing, and pumps? Are you surprised at this number? Why or why not?
4. What would the chromatograph look like if you injected 40 µl of a sample as compared to 20 µl?
5. What would the chromatograph look like if mobile phase A and B were reversed (i.e., beginning with 100% Phase B and increasing Phase A over time)?

REFERENCES

AOAC International. 2000. Method 979.08. Benzoate, Caffeine, and Saccharin in Soda Beverages. *Official Methods of Analysis.* 17th ed. AOAC International, Gaithersburg, MD.

Bridle, P. and Timberlake, F. 1997. Anthocyanins as natural food colours—selected aspects. *Food Chemistry* 58: 103–109.

Hong, V. and Wrolstad, R.E. 1990. Use of HPLC separation/photodiode array detection for characterization of anthocyanins. *Journal of Agricultural and Food Chemistry* 38: 708–715.

Markakis, P. (Ed.). 1982. *Anthocyanins as Food Colors.* Academic Press, New York.

Rounds, M.A., and Nielsen, S.S. 2003. Basic principles of chromatography. Ch. 27, in *Food Analysis*, 3rd ed. S.S. Nielsen (Ed.), Kluwer Academic, New York.

Rounds, M.A., and Gregory, J.F. 2003. High performance liquid chromatography. Ch. 28, in *Food Analysis*, 3rd ed. S.S. Nielsen (Ed.), Kluwer Academic, New York.

Gas Chromatography

Laboratory Developed in part by

Dr Michael Qian,
Oregon State University, Department of Food Science and Technology,
Cornvallis, Oregon

INTRODUCTION

Background

Gas chromatography (GC) has many applications in the analysis of food products. GC has been used for the determination of fatty acids, triglycerides, cholesterol, gases, water, alcohols, pesticides, flavor compounds, and many more. While GC has been used for other food components such as sugars, oligosaccharides, amino acids, peptides, and vitamins, these substances are more suited to analysis by high performance liquid chromatography. GC is ideally suited to the analysis of volatile substances that are thermally stable. Substances such as pesticides and flavor compounds that meet these criteria can be isolated from a food and directly injected into the GC. For compounds that are thermally unstable, too low in volatility, or yield poor chromatographic separation due to polarity, a derivatization step must be done before GC analysis. The two parts of the experiment described here include the analysis of alcohols that require no derivatization step, and the analysis of fatty acids which requires derivatization. The experiments specify the use of capillary columns, but the first experiment includes conditions for a packed column.

Reading Assignment

Reineccius, G.A. 2003. Gas chromatography. Ch. 29, in *Food Analysis*, 3rd ed. S.S. Nielsen (Ed.), Kluwer Academic, New York.

METHOD A: DETERMINATION OF METHANOL AND HIGHER ALCOHOLS IN WINE BY GAS CHROMATOGRAPHY

Introduction

The quantification of higher alcohols, also known as fusel oils, in wine and distilled spirits is important because of the potential flavor impact of these compounds. These higher alcohols include n-propyl alcohol, isobutyl alcohol, and isoamyl alcohol. Some countries have regulations that specify maximum and/or minimum amounts of total higher alcohols in certain alcoholic beverages. Table wine typically contains only low levels of higher alcohols but dessert wines contain higher levels, especially if the wine is fortified with brandy.

Methanol is produced enzymatically during the production of wine. Pectin-methyl-esterase hydrolyzes the methyl ester of α-1,4-D-galacturonopyranose. The action of this enzyme, which is naturally present in grapes and may also be added during vinification, is necessary for proper clarification of the wine. White wines produced in the United States contain

table 18-1   Alcohol Structure and Boiling Point

Alcohol	Structure	b.p.(°C)
Methanol	CH_3OH	64.5
Ethanol	$CH_3—CH_2OH$	78.3
n-Propanol	$CH_3—CH_2—CH_2OH$	97

Isobutyl (2-methyl-1-propanol)	$CH_3—CH—CH_2OH$
	           $\vert$
	        CH_3
		108

Isoamyl (3-methyl-1-butanol)	$CH_3—CH—CH_2CH_2OH$
	           $\vert$
	        CH_3	

Active amyl (2-methyl-1-butanol)	$CH_3—CH_2—CH—CH_2OH$
	              $\vert$
	           CH_3
		128

Benzyl alcohol	⬡—CH_2OH	205

less methanol (4–107 mg/L) compared to red and rosé wines (48–227 mg/L). Methanol has a lower boiling point than the higher alcohols (Table 18-1), so it is more readily volatilized and elutes earlier from a gas chromatography (GC) column.

Methanol and higher alcohols in distilled liquors are readily quantitated by gas chromatography, using an internal standard such as benzyl alcohol, 3-pentanol, or n-butyl alcohol. The method outlined below is similar to AOAC Methods 968.09 and 972.10 [Alcohols (Higher) and Ethyl Acetate in Distilled Liquors].

Objective

Determine the content of methanol, n-propyl alcohol, and isobutyl alcohol in wine by gas chromatography, using benzyl alcohol as the internal standard.

Principle of Method

Gas chromatography uses high temperatures to volatilize compounds that are separated as they pass through the stationary phase of a column and are detected for quantitation.

Chemicals

	CAS No.	Hazards
Benzyl alcohol	100-51-6	Harmful
Ethanol	64-17-5	Highly flammable
Isobutyl alcohol	78-83-1	Irritant
Methanol	67-56-1	Extremely flammable
n-Propyl alcohol	71-23-8	Irritant, highly flammable

Reagents

(**It is recommended that these solutions be prepared by the laboratory assistant before class.)

- Ethanol, 16% (vol/vol) with deionized distilled (dd) water, 100 ml**
- Ethanol, 50% (vol/vol) with dd water, 3100 ml**
- Ethanol, 95% (vol/vol) with dd water, 100 ml**
- Stock Solutions**

  Prepared with known amounts of ethanol and fusel alcohols or methanol:
  1. 10.0 g of methanol and 50% (vol/vol) ethanol to 1000 ml.
  2. 5.0 g of n-propyl alcohol and 50% (vol/vol) ethanol to 1000 ml.
  3. 5.0 g of isobutyl alcohol and 50% (vol/vol) ethanol to 1000 ml.
  4. 5.0 g of benzyl alcohol in 95% (vol/vol) ethanol to 100 ml.
- Working Standard Solutions**

  Prepared from stock solutions, to contain different amounts of each of the fusel alcohols; aliquots of these are used to get standard curves. Prepare four working standards by combining:
  1. 0.5 ml of stock solutions 1, 2, and 3 with 4.5 ml of 50% (vol/vol) ethanol plus 16% (vol/vol) ethanol to 100 ml.
  2. 1.0 ml of stock solutions 1, 2, and 3 with 3.0 ml of 50% (vol/vol) ethanol plus 16% (vol/vol) ethanol to 100 ml.
  3. 1.5 ml of stock solutions 1, 2, and 3 with 1.5 ml of 50% (vol/vol) ethanol plus 16% (vol/vol) ethanol to 100 ml.
  4. 2.0 ml of stock solutions 1, 2, and 3 with 16% (vol/vol) ethanol to 100 ml.

  (*Note*: The final concentration of ethanol in each of these working standard solutions is 18% (vol/vol) ethanol.)

Hazards, Precautions, and Waste Disposal

The alcohols are fire hazards; avoid open flames, breathing vapors, and contact with skin. Otherwise, adhere to normal laboratory safety procedures. Wear safety glasses at all times. Aqueous waste can go down the drain with a water flush.

Supplies

(Used by students)

- Mechanical pipettor, 1000 μl, with tips
- Round bottom flask, 500 ml
- Syringe (for GC)
- 6 Volumetric flasks, 100 ml
- 4 Volumetric flasks, 1000 ml

Equipment

- Analytical balance
- Distillation unit (heating element to fit 500 ml round-bottom flask; cold water condenser)
- Gas chromatography unit:

Column	DB-wax (30 m, 0.32 nm ID, 0.5 um film thickness) (Agilent Technologies, PaloAlto, CA) or equivalent (capillary column), or 80/120 Carbopack BAW/5% Carbowax 20 M, 6 ft × 1/4 in OD × 2 mm ID glass column (packed column)
Injector temperature	200°C
Column temperature	70°C to 170°C@5°C/min
Carrier gas	He at 2 ml/min (N₂ at 20 ml/min for packed column)
Detector	Flame ionization
Attenuation	8 (for all runs)

ID = inner diameter
OD = outer diameter
BAW = base and acid washed

Procedure

(Instructions are given for single standard and sample analysis, but injections can be replicated.)

I. Sample Preparation

1. Fill a 100-ml volumetric flask to volume with the wine sample to be analyzed.
2. Pour the wine into a 500-ml round bottom flask and rinse the volumetric flask several times with dd water to complete the transfer. Add additional water if necessary to bring the volume of sample plus dd water to ca. 150 ml.
3. Distill the sample and recover the distillate in a clean 100-ml volumetric flask. Continue the distillation until the 100-ml volumetric is filled to the mark.
4. Add 1.0 ml of the stock benzyl alcohol solution to 100 ml of each working standard solution and wine sample to be analyzed.

II. Analysis of Sample and Working Standard Solutions

1. Inject 1 μl of each sample and working standard solution in separate runs on the GC column (split ratio 1 : 20). (For packed column, inject 5.0 μl.)
2. Obtain chromatograms and data from integration of peaks.

Data and Calculations

1. Calculate the concentration (mg/L) of methanol, n-propyl alcohol, and isbutyl alcohol in each of the four Working Standard Solutions (see sample calculation below).

Alcohol concentration (mg/L):

Working Standard	Methanol	n-Propyl Alcohol	Isobutyl Alcohol
1			
2			
3			
4			

Example calculations:

Working Standard Solution #1 — contains methanol + n-propyl alcohol + isobutyl alcohol, all in ethanol

Methanol in Stock Solution #1:

$$\frac{10\,\text{g methanol}}{1000\,\text{ml}} = \frac{1\,\text{g}}{100\,\text{ml}} = \frac{0.01\,\text{g}}{\text{ml}}$$

Working Standard Solution #1 contains 0.5 ml of Stock Solution #1.

$$= 0.5\,\text{ml of } 0.01\,\text{g methanol/ml}$$

$$= 0.005\,\text{g methanol} = 5\,\text{mg methanol}$$

That 5 mg methanol is contained in 100 ml volume.

$$= 5\,\text{mg}/100\,\text{ml} = 50\,\text{mg}/1000\,\text{ml}$$

$$= 50\,\text{mg methanol/L}$$

Repeat procedure for each alcohol in each Working Standard Solution.

Peak height ratios for alcohol peaks at various concentrations of methanol, n-propyl alcohol, and isobutyl alcohol, with benzyl alcohol as internal standard:

Alcohol Conc. (mg/L)	Peak Height Ratio[1]		
	Methanol / Benzyl Alcohol	n-Propyl Alcohol / Benzyl Alcohol	Isobutyl Alcohol / Benzyl Alcohol
25			
50			
75			
100			
150			
200			
Wine sample			

[1] Give individual values and the ratio.

2. Calculate the peak height or peak area ratios for methanol, n-propyl alcohol, and isobutyl alcohol, compared to the internal standard, for each of the Working Standard Solutions and the wine sample. To identify which is the methanol, n-propyl alcohol, and isobutyl peak, see the example chromatogram that follows. Note that data from automatic integration of the peaks can be used for these calculations. Report the ratios in a table as shown below. Show an example calculation of concentration for each type of alcohol.

3. Construct standard curves for methanol, n-propyl alcohol, and isobutyl alcohol using the peak height ratios. All lines can be shown on one graph. Determine the equations for the lines.

4. Calculate the peak ratios for methanol, n-propyl alcohol, and isobutyl alcohol in the wine sample, and their concentrations in mg/L.

Questions

1. Explain how this experiment would have differed in standard solutions used, measurements taken, and standard curves used if you had used external standards rather than an internal standard.

2. What are the advantages of using an internal standard rather than external standards for this application, and what were the appropriate criteria to use in selecting the external standard.

METHOD B: PREPARATION OF FATTY ACID METHYL ESTERS (FAMEs), AND DETERMINATION OF FATTY ACID PROFILE OF OILS BY GAS CHROMATOGRAPHY

Introduction

Information about fatty acids profile on food is important for nutrition labeling, which involves the measurement of not only total fat but also saturated, unsaturated, and monounsaturated fat. Gas chromatography is an ideal instrument to determine (qualitatively and quantitatively) fatty acid profile or fatty acid composition of a food product. This usually involves extracting the lipids and analyzing them using capillary gas chromatography. Before such analysis, triacylglycerols and phospholipids are saponified and the fatty acids liberated are esterified to form fatty acid methyl esters (FAMEs) so that the volatility is increased.

Two methods of sample preparation for FAMEs determination will be used in this experiment: (1) sodium methoxide method, and (2) boron trifluoride (BF$_3$) method (AOAC Method 969.33). In the sodium methoxide method, sodium methoxide is used as a catalyst to interesterify fatty acid. This method is applicable to saturated and unsaturated fatty acids containing from 4 to 24 carbon atoms. In the BF_3 method, lipids are saponified, and fatty acids are liberated and esterified in presence of BF_3 catalyst for further analysis. This method is applicable to common animal and vegetable oils and fats, and fatty acids. Lipids that cannot be saponified are not derivatized and, if present in large amount, may interfere with subsequent analysis. This method is not suitable for preparation of methyl esters of fatty acids containing large amounts of epoxy, hydroperoxy, aldehyde, ketone, cyclopropyl, and cyclopentyl groups, and conjugated polyunsaturated and acetylenic compounds because of partial or complete destruction of these groups.

Objective

Utilize two methods to prepare methyl esters from fatty acids in food oils, then determine the fatty acids profile and their concentration in the oils by gas chromatography.

Chemicals

	CAS No.	Hazards
Boron trifluoride (BF$_3$)	7637-07-2	Toxic, highly flammable
Hexane	110-54-3	Harmful, highly flammable, dangerous for the environment
Methanol	67-56-1	Extremely flammable
Sodium chloride (NaCl)	7647-14-5	Irritant
Sodium hydroxide (NaOH)	1310-73-2	Corrosive
Sodium sulfate (Na$_2$SO$_4$)	7757-82-6	Harmful
Sodium methoxide	124-41-4	Toxic, highly flammable

Reagents and Samples

- Boron trifluoride (BF$_3$)—methanol 14% solution
- Hexane (GC grade. If fatty acids contain 20 C atoms or more, heptane is recommended.)
- Methanolic sodium hydroxide 0.5 N (Dissolve 2 g of NaOH in 100 ml of methanol.)
- Oils: pure olive oil, safflower oil, salmon oil
- Reference standard [20A gas–liquid chromatography (GLC) Reference standard, FAME 25 mg is dissolved in 10 ml hexane (Table 18-2) (Nu CHEK Prep, Inc. MN)]

18-2 table **FAME 20A GLC Reference Standard**

No.	Chain	Item	Weight %
1	C14:0	Methyl myristate	2.0
2	C16:0	Methyl palmitate	30.0
3	C16:1	Methyl palmitoleate	3.0
4	C18:0	Methyl stearate	14.0
5	C18:1	Methyl oleate	41.0
6	C18:2	Methyl linoleate	7.0
7	C18:3	Methyl linolenate	3.0

- Sodium methoxide, 0.5 M solution in methanol (Aldrich)
- Sodium chloride, saturated
- Sodium sulfate, anhydrous granular

Hazards, Precautions, and Waste Disposal

Do all work with the boron trifluoride in the hood; avoid contact with skin, eyes, and respiratory tract. Wash all glassware in contact with boron trifluoride immediately after use. Otherwise, adhere to normal laboratory safety procedures. Wear safety glasses at all times. Boron trifluoride, hexane, and sodium methoxide must be disposed of as hazardous wastes. Other wastes likely may be put down the drain using a water rinse, but follow good laboratory practices outlined by environmental health and safety protocols at your institution.

Supplies

(Used by students)

- Boiling flask, 100 ml, with water-cooled condenser for saponification and esterification
- Pasteur pipette
- Syringe
- Vials or sample bottle with tight-seal cap

Equipment

- Analytical balance
- Centrifuge
- Vortex mixer

- Gas chromatography unit (with running conditions):

Instrument	Gas chromatography of HP 5890A GC
Detector	Flame ionization detector
Capillary column	Stabile Wax (Crossbond Carbowax-PEG) (Agilent, CA) or equivalent
Length	30 m
ID (internal diameter)	0.32 mm
Df	1.0 μm
Carrier gas	He
Make-up gas	Nitrogen
Sample Injection	1 μL
Split ratio	1 : 50
Column head pressure	10 psi (70 kPa)
Injector temperature	250°C
Detector temperature	250°C
Temperature program	
Initial oven temperature	160°C
Initial time	2 min
Rate	5°C/min
Final temperature	230°C
Final time	10 min

Procedure

(Instructions are given for single sample preparation and injection, but injections of samples and standards can be replicated.)

I. Preparation of Methyl Esters

Method A: Preparation of Methyl Esters by Boron Trifluoride (adapted from AOAC Method 969.33)

Notes: Methyl ester should be analyzed as soon a possible, or seal in ampule and store in freezer. You might also add equivalent 0.005% 2, 6-di-*tert*-butyl-4-methylphenol (BHT). Sample size needs to be known to determine size of flask and amount of reagents, according to Table 18-3.

1. Add 500 mg sample (see Table 18-3) to 100 ml boiling flask. Add 8 ml methanolic NaOH solution and boiling chip.
2. Attach condenser and reflux until fat globules disappear (about 5–10 min).
3. Add 9 ml BF$_3$ solution through condenser and continue boiling for 2 min.
4. Add 15 ml hexane through condenser and boil for 1 more min.

18-2 table

Determination of Flask Size and Amount of Reagent from Approximate Sample Size

Sample (mg)	Flask (ml)	0.5 N NaOH (ml)	BF$_3$ Reagent (ml)
100–250	50	4	5
250–500	50	6	7
500–750	100	8	9
750–1000	100	10	12

5. Remove the boiling flask and add ca. 15 ml saturated NaCl solution.
6. Stopper flask and shake vigorously for 15 s while solution is still tepid.
7. Add additional saturated NaCl solution to float hexane solution into neck of flask.
8. Transfer 1 ml upper hexane solution into a small bottle and add anhydrous Na$_2$SO$_4$ to remove H$_2$O.

Method B: Preparation of Methyl Esters by Sodium Methoxide Method

1. Using a Pasteur pipette to transfer, weigh 100 mg (± 5 mg) of sample oil to the nearest 0.1 mg into a vial or small bottle with a tight-sealing cap.
2. Add 5 ml of hexane to the vial and vortex briefly to dissolve lipid.
3. Add 250 μl of sodium methoxide reagent, cap the vial tightly, and vortex for 1 min, pausing every 10 s to allow the vortex to collapse.
4. Add 5 ml of saturated NaCl solution to the vial, cap the vial, and shake vigorously for 15 s. Let sit for 10 min.
5. Remove the hexane layer and transfer to a vial containing a small amount of Na$_2$SO$_4$. Do not transfer any interfacial precipitate (if present) or any aqueous phase.
6. Allow the hexane phase containing the methyl esters to sit in contact with Na$_2$SO$_4$ for at least 15 min prior to analysis.
7. Transfer the hexane phase to a vial or small bottle for subsequent GC analysis. (Hexane solution can be stored in the freezer).

II. Injection of Standards and Samples into GC

1. Rinse syringe three times with hexane, and three times with the reference standard mixture (25 mg of 20A GLC Reference Standard FAME dissolved in 10 ml hexane). Inject 1 μl of standard solution, remove syringe from injection port, then press start button. Rinse syringe

again three times with solvent. Use the chromatogram obtained as described below.

2. Rinse syringe three times with hexane, and three times with the sample solution prepared by Method A. Inject 1 μl of sample solution, remove syringe from injection port, then press start button. Rinse syringe again three times with solvent. Use the chromatogram obtained as described below.

3. Repeat Step 3 for sample solution prepared by Method B.

Data and Calculations

1. Report retention times and relative peak areas for the peaks in the chromatogram from the FAME reference standard mixture. Use this information to identify the seven peaks in the chromatogram.

Peak	Retention Time	Peak Area	Identity of Peak
1			
2			
3			
4			
5			
6			
7			

2. Using the retention times for peaks in the chromatogram from the FAME reference standard mixture, and your knowledge of the profile of the oil, identify the peaks in the chromatograms for each type of oil analyzed. [Cite your source(s) of information on the fatty acid profile of each oil.] Report results for samples from both methods of derivatization.

Results from chromatograms using boron trifluoride method to prepare methyl esters:

	Safflower Oil		Pure Olive Oil		Salmon Oil	
Peak	Retention Time	Identity	Retention Time	Identity	Retention Time	Identity
1						
2						
3						
4						
5						
6						
7						

Results from chromatograms using sodium methoxide method to prepare methyl esters:

	Safflower Oil		Pure Olive Oil		Salmon Oil	
Peak	Retention Time	Identity	Retention Time	Identity	Retention Time	Identity
1						
2						
3						
4						
5						
6						
7						

3. For the one oil analyzed by your group, prepare a table (with appropriate units) comparing your experimentally determined fatty acid profile to that found in your cited literature source.

	Quantity Determined	
Quantity in Literature	Boron Trifluoride Method	Sodium Methoxide Method
C14:0		
C16:0		
C16:1		
C18:0		
C18:1		
C18:2		

Type of oil tested:

Questions

1. Comment on the similarities and differences in the fatty acid profiles in question 3 of Data and Calculations, comparing experimental data to literature reports. From the results, compare and decide which method of esterification to obtain FAMEs was better for your sample.

2. The approach taken in this lab provides a fatty acid profile for the oils analyzed. This is sufficient for most analytical questions regarding fatty acids. However, determining the fatty acid profile is not quite the same thing as quantifying the fatty acids in the oil. (Imagine that you wanted to use the results of your GC analysis to calculate the amount of mono- and polyunsaturated fatty acids as grams per a specified serving size of the oil.) To make this procedure sufficiently quantitative for a purpose like that just described, an internal standard must be used.

   a. Why is the fatty acid profiling method used in lab inadequate to quantify the fatty acids?

   b. What are the characteristics required of a suitable internal standard for FAME quantification by GC and how does this overcome the problem(s) identified in (a)?

c. Would the internal standard be added to the reference standard mixture and the sample, or only to one of these?

d. When would the internal standard be added?

REFERENCES

Amerine, M.A., and Ough, C.S. 1980. *Methods for Analysis of Musts and Wine*. John Wiley & Sons, New York.

AOAC International. 2000. Methods 968.09, 969.33, and 972.10. *Official Methods of Analysis*, 17th ed. AOAC International, Gaithersburg, MD.

Martin, G.E., Burggraff, J.M., Dyer, R.H., and Buscemi, P.C. 1981. Gas–liquid chromatographic determination of congeners in alcoholic products with confirmation by gas chromatography/mass spectrometry. *Journal of the Association of Official Analytical Chemists* 64: 186.

Pike, O.A. 2003. Fat characterization, Ch. 14, in *Food Analysis*, 3rd ed. Kluwer Publishers, New York.

Reineccius, G.A. 2003. Gas chromatography. Ch. 29, in *Food Analysis*, 3rd ed. S.S. Nielsen (Ed.); Kluwer Academic, New York.

19
chapter

Viscosity Measurement Using a Brookfield Viscometer

Laboratory Developed by
Dr Christopher R. Daubert and Dr Brian E. Farkas
North Carolina State University, Department of Food Science,
Raleigh, North Carolina

INTRODUCTION

Background

Whether working in product development, quality control, or process design and scale-up, rheology plays an integral role in manufacturing the best products. Rheology is a science based on fundamental physical relationships concerned with how all materials respond to applied forces or deformations.

Determination and control of the flow properties of fluid foods is critical for optimizing processing conditions and obtaining the desired beneficial effects for the consumer. Transportation of fluids (pumping) from one location to another requires pumps, piping, and fittings such as valves, elbows, and tees. Proper sizing of this equipment depends on a number of elements but primarily on the flow properties of the product. For example, the equipment used to pump a dough mixture would be very different from that used for milk. Additionally, rheological properties are fundamental to many aspects of food safety. During continuous thermal processing of fluid foods the amount of time the food is in the system (known as the residence time or RT), and therefore the amount of heating or "thermal dose" received, directly relates to its flow properties.

The rheological properties of a fluid are a function of composition, temperature, and other processing conditions. Identifying how these parameters influence flow properties may be performed using a variety of rheometers. In this laboratory we will measure the viscosity of three liquid foods using Brookfield rotational viscometers—common rheological instruments widely used throughout the food industry.

Reading Assignment

Daubert, C.R., and Foegeding, E.A. 2003. Rheological principles for food analysis. Ch. 30, in *Food Analysis*, 3rd ed. S.S. Nielsen (Ed.), Kluwer Academic, New York.

Singh, R.P., and Heldman, D.R. 2001. *Introduction to Food Engineering*, 3rd ed., pp. 69–78, 144–157. Academic Press, San Diego, CA.

Objectives

1. Become familiar with the study of fluid rheology;
2. Gain experience measuring fluid viscosity; and
3. Observe temperature and (shear) speed effects on viscosity.

Supplies

- 3 beakers, 250 ml
- French salad dressing
- Honey
- Thermometer

Equipment

- Brookfield rotational viscometer model LV and spindle #3
- Refrigerator

PROCEDURE

1. Prior to evaluating the samples, make sure the viscometer is level. Use the leveling ball and circle on the viscometer.
2. Fill a beaker with 200 ml honey and the two remaining beakers with 200 ml salad dressing. Place one of the beakers of salad dressing in a refrigerator 1 hr prior to analysis. The remaining beakers shall be allowed to equilibrate to room temperature.
3. Because rheological properties are strongly dependent on temperature, measure and record fluid temperatures prior to each measurement.
4. On the data sheet provided, record the viscometer model number and spindle size, product information (type and brand, etc.) and the sample temperature.
5. Immerse the spindle into the test fluid (i.e., honey, salad dressing) up to the notch cut in the shaft; the viscometer motor should be off.
6. Zero the digital viscometers if necessary.
7. Set the motor at the lowest speed (revolutions per minute—rpm) setting. Once the digital display shows a stable value, record the percentage of full scale torque reading. Increase the rpm setting to the next speed and again record the percentage of full-scale torque reading. Repeat this procedure until the maximum rpm setting has been reached *or* 100% (but not higher) of the full-scale torque reading is obtained.
8. Stop the motor and slowly raise the spindle from the sample. Remove the spindle and clean with soap and water, then dry.
9. A factor exists for each spindle-speed combination (Table 19-1).

19-1
table

Factors for Brookfield Model LV (Spindle #3)

Speed (rpm)	Factor
0.3	4000
0.6	2000
1.5	800
3	400
6	200
12	100
30	40
60	20

For every dial reading (percentage full-scale torque), multiply the display value by the corresponding factor to calculate the viscosity with units of mPa-s.

Example:

A French salad dressing was tested with a Brookfield LV viscometer equipped with spindle #3. At a speed of 6 rpm, the display read 40.6%. For these conditions, a viscosity is calculated:

$$\eta = 40.6 \times 200 = 8120 \text{ mPa-s} = 8.12 \text{ Pa-s}$$

10. Repeat Steps 3–9 to test all samples.
11. Once all the data has been collected for the salad dressing and honey, remove the salad dressing sample from the refrigerator and run the same procedure. Be sure to record the sample temperature.
12. You may choose to run the samples in duplicate or perhaps triplicate. Data from samples collected under identical conditions may be pooled to generate an average reading.

DATA

Date:
Product Information:
Viscometer Make and Model:
Spindle Size:

Spindle Speed (rpm)	% Reading	Factor	Viscosity (mPa-s)

CALCULATIONS

1. Sketch and describe (label) the experimental apparatus.

2. Calculate the viscosity of the test fluids at each rpm.
3. Plot viscosity versus rpm for each fluid on a single graph.
4. Label the plots with the type of fluid based on the response of viscosity to speed (rpm). *Keep in mind, the speed is proportional to the shear rate. In other words, as the speed is doubled, the shear rate is doubled.*

Questions

1. What is viscosity?
2. What is a Newtonian fluid? What is a non-Newtonian fluid? Which of your materials responded as a Newtonian fluid?
3. What effect does temperature have on the viscosity of fluid foods?
4. How may food composition impact the viscosity? What ingredient in the salad dressing may impart deviations from Newtonian behavior?
5. Describe the importance of viscosity in food processing, quality control, and consumer satisfaction.
6. For samples at similar temperatures and identical speeds, was the viscosity of honey ever less than the viscosity of salad dressing? Is this behavior representative of the sample rheology at all speeds?
7. Why is it important to test samples at more than 1 speed?

REFERENCES

Daubert, C.R., and Foegeding, E.A. 2003. Rheological principles for food analysis. Ch. 30, in *Food Analysis*, 3rd ed. S.S. Nielsen (Ed.), Kluwer Academic, New York.

Singh, R.P., and Heldman, D.R. 2001. *Introduction to Food Engineering*, 3rd ed., pp. 69–78, 144–157. Academic Press, San Diego, CA.